Research Trends in Air Pollution Control: Scrubbing, Hot Gas Clean-up, Sampling and Analysis

R. Mahalingham, and Alfred J. Engel, editors

Theodore G. Brna
Stuart W. Churchill
Paul T. Cunningham
A. S. Damle
David L. Etherton
Dale A. Furlong
T. L. Holmes
Stanley A. Johnson
John J. Kalvinskas
Samir P. Kothari
Romesh Kumar

Noam Lior
R. Mahalingham
Chat P. Mohan
Ronald L. Ostop
Avinash N. Patkar
D. T. Pratt
Alan D. Randolph
Naresh Rohatgi
Eric A. Samuel
J. Thomas Schrodt
S.-K. Tang

B. K. Thota

AIChE Symposium Series

Number 211   1981   Volume 77

Published by

American Institute of Chemical Engineers

345 East 47 Street   New York, N. Y. 10017

Copyright 1981

American Institute of Chemical Engineers
345 East 47 Street, New York, N.Y. 10017

Library of Congress Cataloging in Publication Data
Main entry under title:

Research trends in air pollution control.

(AIChE symposium series ; no. 211)
1. Flue gases—Purification—Addresses, essays, lectures. I. Mahlingam, R., 1938-   . II. Engel, Alfred J.
III. Series.

TD885.R47   628.5'32   81-19133
ISSN 0065-8812   AACR2
                                         ISBN 0-8169-0219-4

Printed in the United States of America by
Lew A. Cummings Co., Inc.

FOREWORD

Chemical engineers and other professionals have a key role to play in the preservation of the environment, in addition to economic and efficient operation of process plants. Additionally, they have the responsibility of recycling the key components from the effluent streams, wherever feasible, in order to conserve our valuable resources.

The theme of the AIChE 89th National Meeting held in Portland, Oregon, in August 1980 was "Use of Renewable and Non-Renewable Resources—A Challenge for the Future". Five symposia in the Air area were held at this meeting to reflect the above theme. The sessions were Air Pollution Transport, Hot Gas Clean-Up (2 sessions), Air Pollution Sampling and Analysis, and Recent Trends in Control Practices for Industrial Emissions. At the 73rd Annual AIChE Meeting in Chicago in November 1980, no specific sessions dealing with air pollution were held, but four sessions on coal research and coal use were presented. Between these two AIChE meetings, a total of 59 papers were presented. Some of these have already appeared in *Chemical Engineering Progress* and some others are scheduled to appear in the AIChE Symposium Series on coal. Out of the remaining papers, those printed here represent a wide diversity of interest and should be of lasting value. No doubt there were other equally qualified papers presented that could not be included here because of space limitations.

The editors hope that the papers in this volume will be a resource material for a long time and will be useful in opening up investigations along new avenues.

In addition to the various session chairmen and co-chairmen who deserve our thanks, one person who deserves special thanks here is Charles A. Brown, a graduate student at Washington State University and presently a graduate student at the University of Washington.

R. Mahalingam, *editor*
Washington State University
Pullman, Washington

A. J. Engel, *editor*
Pennsylvania State University
University Park, Pennsylvania

CONTENTS

FUEL GAS DESULFURIZATION IN FLUIDIZED-BEDS OF GASIFIER WASTE ASHES

J. THOMAS SCHRODT

Department of Chemical Engineering
University of Kentucky
Lexington, Kentucky 40506

CHAT P. MOHAN

Monsanto Chemical Company
St. Louis, Missouri

Gasifier ashes, inherently containing iron oxides, show excellent high temperature reactivity for the H_2S, COS, and CS_2 found in low energy coal gases. Data gathered from small fluid beds of four selected ashes contacted with synthesized gases containing H_2S support the good performance and efficiency of his desulfurization process. Solid phase diffusion of ionic iron in the particles, carbon deposition, ash regeneration, and secondary process reactions are discussed.

A significant fraction of the sulfur in coal appears in the pyrite (FeS_2) and sulfate ($FeSO_4$) forms. During coal gasification the iron is oxidized to hematite (Fe_2O_3) and magnetite (Fe_3O_4) and rejected as part of the bottoms ash. The sulfur is converted to hydrogen sulfide (H_2S) and lesser amounts of carbon disulfide (CS_2) and carbonyl sulfide (COS). Although these sulfur compounds account for less than one mole percent of the product fuel gas, they must be removed before combustion for environmental reasons and to prevent corrosion to process and combustion equipment. If desulfurization is carried out while the fuel gas is still at gasifier exit conditions, i.e. 800° to 1200°K and 1 to 100 atmospheres pressure, 5 to 10 percent of the heating value of the fuel can be retained in the form of sensible heat. These facts provide incentive for development of new high temperature desulfurization processes. Murthy, et al. (1977), Edwards (1979), Morrison (1979), Schrodt and Hahn (1976), and Mohan (1980) have reviewed the most recent developments in this field of hot gas cleanup.

Our objective in this work is to determine if gasifier bottoms ashes containing iron oxides can effectively remove hydrogen sulfide from hot, made-up fuel gases, when the latter are passed through small fluidized-beds of the ashes. We seek to identify through X-Ray and Energy Dispersive Analyses of the solids, and

influent and effluent gas analyses, the gas, and gas-solid reactions, and the effects of several selected parameters on these. Through careful reduction of the effluent gas concentration data we seek to identify the rate controlling reaction steps, and to develop and verify global, fluid-bed models.

EXPERIMENTAL

Three gasifier ashes, containing 5 to 23 weight percent Fe_2O_3 were selected for testing as the solid reactants. These were crushed and three U.S.-Mesh sized fractions: 70/80, 80/120, and 120/170, were prepared. The chemical, physical, and fine particle characteristics of these are reported in the dissertation of the junior author (Mohan, 1980).

The desulfurization tests were carried out at atmospheric pressure in a 20-mm x 100-mm quartz tube with a porous, quartz disc fused to the inner wall. This disc supported the ash, which was fluidized to about 2X the minimum fluidization velocity by a synthesized fuel gas of nominal composition: 20% CO, 10% H_2, 17% CO_2, 3% CH_4, 49% N_2, and the remainder H_2S (0.5 to 1.0%). Gas compositions were measured by a dual column GC equipped with a TC detector and an on-line CDS. The gases were preheated before contacting the ash bed. The quartz tube was located in a ±1.0°K controlled high-temperature furnace and temperatures were monitored via quartz-sheathed chromel-alumel thermocouples.

Before and during the reaction studies, the fluidizing and decrepitation behavior of the ashes was observed both at room temperature, and at 756°, 811°, 922°, and 1033°K. No significant decrepitation of any of the four ashes in the three sized fractions was measured after 10 days of fluidization at room temperature nor after 60 days at the higher temperatures. At 1033°K ash particles tended to adhere to one another and to the wall of the tube. Continuous, uniform fluidization at this high temperature was judged difficult. At 2X the minimum fluidization velocity, the bed showed an expansion of about 1.8, with good solid mixing.

RESULTS AND DISCUSSION

Desulfurization started at the entrance of the sorbent bed and progressed under a concentration wave of H_2S that moved very slowly within a reaction zone through the bed. In front of the reaction zone, where $H_2/H_2S \gg 100$ and $CO/CO_2 \cong 1.6$, Fe_2O_3 in the ash was rapidly reduced to Fe_3O_4 and sometimes even to $Fe_{.95}O$ (wüstite):

$$3Fe_2O_3 + H_2 \rightarrow 2Fe_3O_4 + H_2O$$
$$K_{1000°K} = 1.0 \times 10^5 \qquad (1)$$

$$3Fe_2O_3 + CO \rightarrow 2Fe_3O_4 + CO_2$$
$$K_{1000°K} = 1.1 \times 10^5 \qquad (2)$$

$$Fe_3O_4 + H_2 \rightarrow 3FeO + H_2O$$
$$K_{1000°K} = 0.34 \qquad (3)$$

$$Fe_3O_4 + CO \rightarrow 3FeO + CO_2$$
$$K_{1000°K} = 0.63 \qquad (4)$$

Powder X-Ray scans on fresh, and pretreated ash recovered from the reactor after a brief exposure to the reducing gases are shown in Figure 1. The diffraction patterns of these two scans strongly support the occurrence of reactions (1) and (2). Sharp increases and decreases in the initial concentrations of CO_2 and H_2O; and CO and H_2 respectively were also noted, and support the reduction reactions. The Fe-O-S phase diagram presented in Figure 2 indicates a thermodynamic correctness to the above conclusions.

It was also noted that CO, CO_2, H_2, and H_2O reached near equilibrium concentrations relative to the water-gas-shift reaction:

$$CO + H_2O \rightarrow CO + H_2 \quad K_{1000°K} = 1.75 \qquad (5)$$

Product to reactant concentration ratios for this reaction were 2.7 and 2.0 at 922° and 1033°K, while the corresponding theoretical values of K are 2.2 and 1.5, respectively. At lower temperatures, reaction rates were so low that apparently equilibrium could not be attained. Evans (1975) measured reaction rates for the reverse water-gas-shift reaction, homogeneously, and over reduced, and sulfided ashes and concluded the reaction is catalyzed by minerals in the ash. Surprisingly, sulfidation of the magnetite did not affect the reaction rate.

As the fuel gas entered the fluidized-bed, reaction commenced between H_2S and Fe_3O_4 and/or $Fe_{.95}O$:

$$Fe_3O_4 + 3H_2S + H_2 \rightarrow 3FeS + 4H_2O$$
$$K_{1000°K} = 2.67 \times 10^7 \qquad (6)$$

$$FeO + H_2S \rightarrow FeS + H_2O$$
$$K_{1000°K} = 4.21 \times 10^2 \qquad (7)$$

Large equilibrium constants favor complete removal of the sulfides from the gas. The solid sulfide is actually pyrrhotite, $Fe_{.947}O$. X-Ray analyses shown in Figure 1 of fully sulfided ashes indicate pyrrhotite is the only iron sulfide formed during desulfurization. This finding is contrary to results reported by Schultz and Berber (1970) but in agreement with those of Joshi and Leuenberger (1977). According to the phase diagram shown in Figure 2, this is the only stable solid sulfide that could reside in the bed at the prevailing conditions. At temperatures below 1000°K the H_2/H_2S ratio has to be less than 6.5×10^{-3} to have pyrite as the stable sulfide; this concentration ratio is unlikely to occur during fuel gas processing. For long on-stream times there was no detectable gaseous sulfide in the effluents. The concentration wave of H_2S was formed near the inlet and moved through the bed at a rate dependent upon the availability of unreacted iron and hydrogen sulfide.

As the availability of reactant iron oxide in the bed diminished, measurable and increasing concentrations of two sulfide gases: carbonyl sulfide (COS) and H_2S broke through the bed. In Figure 3, which shows this result, it should be noted that the sum of the steady-state exit concentrations of COS and H_2S equals the inlet concentration of H_2S. The presence of the COS is a result of dehydrogenation of H_2S according to the equation:

$$H_2S + CO_2 \rightarrow COS + H_2O \quad K_{1000°K} = 0.033 \quad (8)$$

At all temperatures investigated reaction (8) was found to be at equilibrium. An identical result was also reported by Schrodt (1978) in some earlier fixed-bed studies on this system.

When fully sulfided ashes were fluidized with streams of 3 percent O_2 in N_2, the sulfur in the pyrrhotite was released as SO_2:

$$2FeS + 7/2 \ O_2 \rightarrow Fe_2O_3 + 2SO_2$$
$$K_{1000°K} = 10^{90} \quad (9)$$

Unlike the earlier fixed-bed studies, no significant amount of elemental sulfur was formed.

Total ash sulfur capacities are evaluated by the equation:

$$M_o = \frac{QC_{AO}}{q} \int_o^{t_E} (1 - \frac{C_A}{C_{AO}}) \ dt \quad (10)$$

where t_E = time when $C_A/C_{AO} \rightarrow 1.0$, and sorption bed efficiencies by the equation:

$$\tau_{0.1} = \frac{QC_{AO}t_{0.1}}{M_o q} \quad (11)$$

where here $t_{0.1}$ = time when $C_A/C_{AO} \rightarrow 0.1$. Tau is a dimensionless time. M_o and $\tau_{0.1}$ were parameters used to evaluate the effects of the independent variables on the desulfurization reaction rates. These effects were examined from plots of C_A/C_{AO} versus τ.

M_o was expected to be a function of the inherent iron in the ashes only; however, it was discovered that fresh ashes - those which had not been subjected to a series of sorption-regeneration cycles - had sulfur capacities that were only a fraction of their theoretical capacity. As the number of cyclic uses was increased, as illustrated in Figure 4, the values of M_o and $\tau_{0.1}$ also increased and reached constant values after 6 to 10 cycles. With a Western Kentucky No. 9 ash containing 22.13 weight percent Fe_2O_3, a stabilized value of $M_o = 1.17 \times 10^{-3}$ g mol/g ash and $\tau_{0.1} = 0.835$ was reached at 922°K. When the desulfurization temperature was suddenly decreased from 922° to 810°K, the values of M_o and $\tau_{0.1}$ also decreased, as shown in Figure 4, and stabilized at lower values.

Analyses of random particles of ash by SEM and EDAX show that the cyclic phenomenon is related to a diffusion of iron from within the core of the particles to the surface structure of the pores as indicated in Figure 5. Figure 6 shows the reproducibility of the fluidized-bed data after stabilization of the ash.

Additional data show that ash capacity and efficiency increase with temperature (Figure 7), Fe_2O_3 concentration, and decreasing fluid velocity. Some selected results are presented in Table 1.

MATHEMATICAL MODEL

The reaction of H_2S with Fe_3O_4 within ash particles involves the following transport and kinetic steps:

- Diffusion of H_2S from the bulk gas phase to the solid particle.
- Diffusion of H_2S through interstices of the particle to reaction sites.
- Chemical reaction of H_2S with unreacted Fe_3O_4.
- Counter-diffusion of the product gas, H_2O, out of the solid.

SEM analyses showed the oxide is concentrated after cyclic use in a thin layer near the particle surface. The layers approximate flat plates of thickness 10 to 50 μ fixed by the parent iron oxide concentrations. Two asymptotic regimes of reaction control are suggested: a region of chemical control where Fe_3O_4 grains react in a spatially uniform way with a uniform concentration of H_2S, and a region of diffusion control where reactant gas diffuses through interstices and reacts rapidly at a sharp solid product-reactant interface.

Unlike the previously cited fixed-bed desulfurization studies in which the concentration breakthrough curves were mirror images of the inner bed, reaction concentrations, the breakthrough curves from the fluidized-bed studies were more logically images of the final stages of reaction of all the solid particles in the bed. Therefore, when the effluent concentration data were applied to global reaction rate models, the parameters deduced are at best representative of the final stages of reaction.

A uniform chemical reaction rate model describes the sulfide breakthrough data test. When material balances were developed for the sulfur components in both the gas the solid phases with the basic assumptions of (1) isothermal plug-flow, (2) first order gas kinetics, (3) irreversible reaction, (4) pseudo steady-state relative to bed and particles, (5) perfect solid mixing, and (6) uniform dispersal of H_2S in the solid, the following coupled dimensionless equations were obtained:

$$\frac{d\overline{Y}}{d\overline{Z}} = \frac{\overline{Y}\overline{X}^m}{N_K + N_F\overline{X}^m} \qquad (12)$$

$$\frac{d\overline{X}}{d\tau} = 1 - \exp\left[\frac{-\overline{X}^m}{N_K + N_F\overline{X}^m}\right] \qquad (13)$$

where

$$N_K \equiv \frac{u}{Lk_vW_o^m\rho_B} \qquad (14)$$

$$N_F \equiv \frac{uR}{3Lk_m(1-\varepsilon)} \qquad (15)$$

$$\overline{Y} \equiv \frac{C_A}{C_{AO}}, \quad \overline{X} \equiv \frac{W}{W_o}, \quad \overline{Z} \equiv \frac{1}{L}$$

To estimate the mass transfer parameter K_F, a value for the mass transfer coefficient k_m was necessary. The following correlation from Kunii and Levenspiel (1969) was selected:

$$\frac{k_md_p}{D_m} = 0.374 \, Re^{1.18} \qquad (16)$$

For all test runs $Re \cong 1.0$, $k_m \cong 1.0$ cm/sec and $K_m \cong 0.01$ to 0.02.

Equations (12) and (13) were coded for Fourth Order Runge-Kutta solution. The parameters to be determined were m, the exponent on the solid reactant concentration, and N_K, the reaction rate parameter. Through trial and error procedure, curves generated for m = 2 were found to best match the breakthrough curves of 38 test runs. A similar procedure was used to arrive at values of N_K and from these the kinetic rate constants, k_v was calculated. These were statistically analyzed to check for the effects of the operating variables. The temperature effect was expressed by the Arrhenius relationship and the following correlation was obtained:

$$k_v = k_o \, C_{AO}^{'a} \left(\frac{u}{u_{mf}}\right)^b e^{-E/RT} \qquad (17)$$

Values of the parameters are listed here:

$$k_o = 2.49 \times 10^{13} \, \frac{g \, ash\text{-}cm^3}{(gmol \, Fe_2O_3)^2 \, min}$$

$$E = 10{,}882 \, cal/gmol$$

$$a = -1.48$$

$$b = -0.788$$

The proper rate expression is

$$-R_A = k_vC_AW^2 \qquad (18)$$

Figure 8 shows a predicted sulfide breakthrough curve and the declining concentration of Fe_2O_3 within the fluidized-bed reactor for one run with actual data superimposed.

It is believed that this work will help to support the development of a continuous process for desulfurizing hot coal-derived fuel gases.

ACKNOWLEDGEMENT

This work was supported by the U.S. Department of Energy under contract E-(40-1)-5076 and the Commonwealth of Kentucky through the Institute for Mining and Minerals Research.

NOMENCLATURE

C_A	Bulk concentration of gas reactant, $gmol/cm^3$
C_A'	Bulk concentration of gas reactant, mole %
D_m	Diffusivity of gas reactant in fuel, cm^2/sec
d_p	Mean particle diameter, cm
E	Activation energy, cal/gmol
K	Thermodynamic equilibrium constant
k_m	Mass transfer coefficient, cm/sec
k_v	Reaction rate constant, g ash-$cm^3/(g$ mol $Fe_2O_3)^2$ min
L	Length of fluidized bed, cm
l	Position in bed, cm
M	Ash sorbate capacity, g mol/g ash
N_F	Parameter defined by Eq. (15)
N_K	Parameter defined by Eq. (14)
Q	Total flow rate of gas, cm^3/min
q	Ash load in bed, g
R	Gas law constant, cal/g mol °K
R_A	Reaction rate
Re	Reynolds number, $d_p u \rho / \mu$
T	Temperature, °K
t	Time, min
u	Fluid velocity, cm/min
W	Solid reactant concentration, g mol/g ash
$\overline{X}$	Dimensionless solid concentration
$\overline{Y}$	Dimensionless gas concentration
$\overline{Z}$	Dimensionless position in bed

GREEK LETTERS

ε	Porosity of bed
ρ_B	Bulk density of bed, g/cm^3
τ	Dimensionless time
$\tau_{0.1}$	Sorption efficiency

SUBSCRIPT

o	Initial value

LITERATURE CITED

M. S. Edwards, "H2S-Removal Processes for Low-BTU Gas", Oak Ridge National Laboratory Report TM-6077, January 1979.

J. F. Evans, "Secondary Factors in the Hot Ash Desulfurization Process", M.S. Thesis, University of Kentucky, 1978.

D. K. Joshi and E. L. Leuenberger, "Hot Low BTU Producer Gas Desulfurization in Fixed Bed of Iron Oxide-Hy Ash", U.S. Department of Energy Report Fe-2033-19, September 1977.

D. Kunii and O. Levenspiel, "Fluidization Engineering", p. 200, John Wiley & Son, New York, 1969.

C. P. Mohan, "Desulfurization of Fuel Gases in Fluidized Beds of Gasifier Waste Ashes", Ph.D. Dissertation, University of Kentucky, December 1980.

G. F. Morrison, "Hot Gas Cleanup", IEA Coal Research Report ICTIS/TR 03, March 1979.

B. N. Murthy, et al., "Fuel Gas Cleanup Technology for Coal Gasification", U.S. Department of Energy Report FE-2220-15, March 1977.

J. T. Schrodt and O. J. Hahn, "Hot Fuel Gas Desulfurization", Report IMMR 15-PDII-76, University of Kentucky, May 1976.

J. T. Schrodt, "Hot Gas Desulfurization with Gasifier Ash Sorbents", Proceedings of the Symposium on Potential Health and Environmental Effects of Synthetic Fossil Fuel Technologies, 32, July 1979.

F. G. Schultz and J. S. Berber, "Hydrogen Sulfide Removal from Hot Producer Gas with Sintered Absorbents", Jour. of Air Pollution Control Assoc., 20, 93, 1970.

TABLE 1. Selected Desulfurization Results

Run	Ash	Fe_2O_3 (wt%)	Particle Size cm	Temperature °K	Fluid Velocity cm/sec	Sulfur Capacity of Ash g/100g	$\tau_{0.1}$
061	W.Ky. 9	22.13	0.0151	1033	8.11	3.98	0.865
049	W.Ky. 9	22.13	0.0151	922	7.26	3.74	0.833
053	W.Ky. 9	22.13	0.0151	811	6.52	3.17	0.705
058	W.Ky. 9	22.13	0.0151	756	6.07	3.19	0.663
075	W.Ky. 9	22.13	0.0107	811	5.58	4.12	0.726
126	W.Ky. 9	22.13	0.0151	811	9.55	4.21	0.700
095	Elkhorn 3	9.51	0.01935	811	6.80	1.29	0.550
111	Rosebud	8.19	0.01935	811	6.80	1.07	0.630

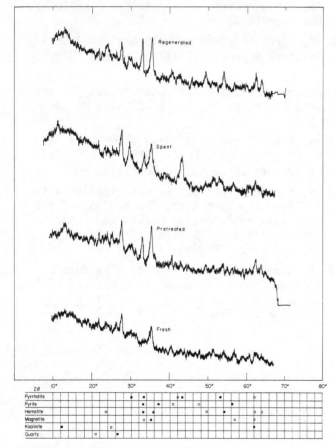

Figure 1. X-ray analyses of fresh, pretreated, spent and regenerated Western Ky. No. 9 coal ash. (■ primary and □ secondary peaks)

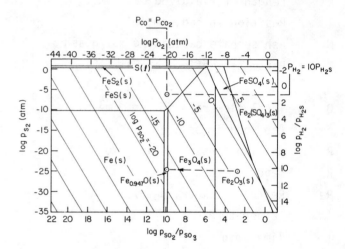

Figure 2. Thermochemistry of iron-sulfur-oxygen at 1000°K.

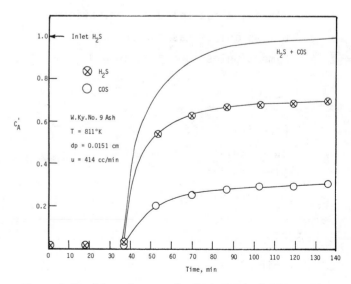

Figure 3. Breakthrough curves for COS, H_2S, and COS + H_2S.

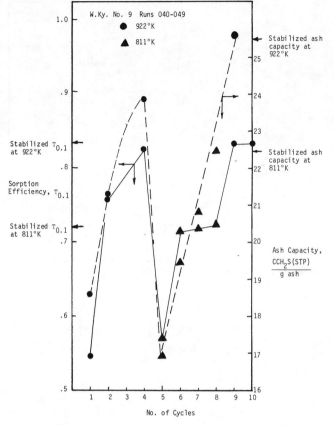

Figure 4. Ash sulfur capacities and sorption efficiencies as a function of cyclic use.

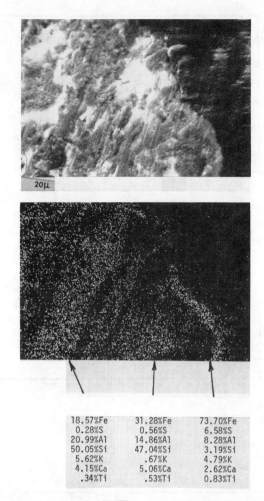

18.57%Fe	31.28%Fe	73.70%Fe
0.28%S	0.56%S	6.58%S
20.99%Al	14.86%Al	8.28%Al
50.05%Si	47.04%Si	3.19%Si
5.62%K	.67%K	4.79%K
4.15%Ca	5.06%Ca	2.62%Ca
.34%Ti	.53%Ti	0.83%Ti

Figure 5. Scanning electron micrograph of cross section of ash particle after ten regenerations with air. Right hand picture shows qualitative distribution of iron. EDAX results indicate high iron concentrations along edge of particle with residual sulfur.

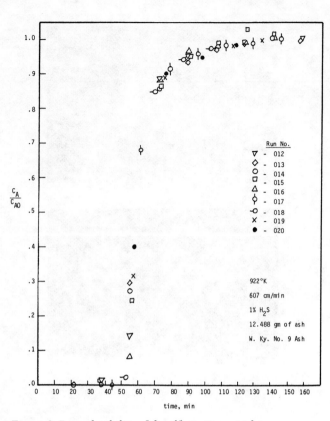

Figure 6. Reproducibility of desulfurization rate data.

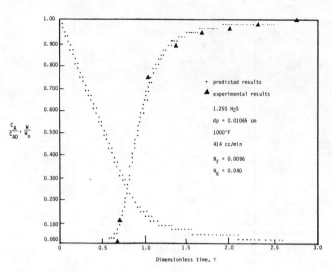

Figure 8. Dimensionless breakthrough curves: predicted by mathematical model and actual data.

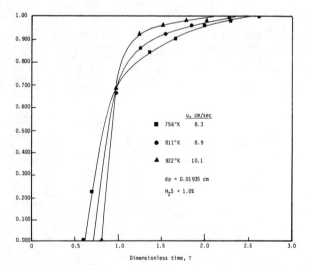

Figure 7. Presentation of rate data in dimensionless form to show effect of temperature.

CATALYSTS DEVELOPMENT AND EVALUATION IN THE CONTROL OF HIGH-TEMPERATURE NO$_x$ EMISSIONS

R. MAHALIGAM

B. K. THOTA

and

D. T. PRATT

Department of Chemical Engineering
Washington State University
Pullman, Washington, 99164

Supported catalysts, which could effectively reduce NO$_x$ emissions from high temperature sources, were developed and evaluated. The variables studied were the inlet concentration of NO$_x$, the gas hourly space velocity, and the effect of the presence of a reducing agent.

Experimental work involved the passing of gas mixtures containing NO$_x$ through a heated catalyst bed 10 cm long and contained inside a 2.7 cm I.D. ceramic tube. In selecting the catalyst, particular attention was paid to its stability under the given operating conditions (about 1000°C and slightly above atmospheric pressures) for possible placement in the immediate vicinity of the combustion zone. Supported nickel and supported cobalt oxide catalysts have been studied. Reaction section inlet and exit gas streams were analyzed for NO$_x$ concentrations with a chemiluminescent NO/NO$_x$ analyzer.

Results from this work show that the supported nickel catalyst is relatively more effective for NO$_x$ decomposition. When, however, carbon monoxide in excess was introduced into the reaction gases, chemical reduction approaching 100 percent NO$_x$ removal on both supported nickel and supported cobalt catalysts was observed.

Oxides of nitrogen have been identified as toxic air pollutants and efforts are being directed towards alleviating the problem by controlling their emissions. Out of a total NO$_x$ emission of 20 million tons per year, combustion of fuel is by far the largest stationary and mobile source of NO$_x$ ([1]). 55% of all NO$_x$ emissions originate from stationary combustion sources, 40% from mobile sources, and the rest from chemical process industries, etc. The combustion sources are boilers, internal combustion engines, gas turbines, and incinerators. Gas turbines in electric utilities ([2]) and in heavy-duty vehicular applications are expected to grow at phenomenal rates. New Source Performance Standards (NSPS) for NO$_x$ are expected to be tightened in future years, especially with increase in concern over secondary particulate formation and acid rain ([3]).

Extensive research has been and is currently being carried out to cope with the problem of NO$_x$ emission from different sources. Very few publications, if any, have come out dealing with NO$_x$ emission problems arising from high temperature sources. Con-

ventional methods applicable to automobile exhaust emission control may not work or may need modifications for situations where high temperatures and large volumes of exhaust gases occur.

BACKGROUND

Formation of NO$_x$

NO$_x$ refers to the sum of nitric oxide (NO) and NO$_2$. Oxides evolved from the nitrogen present in the fuel are called fuel NO$_x$ while those that evolve from the nitrogen in the combustion air are termed thermal NO$_x$.

Nitric oxide formation in combustion processes proceeds according to the following reactions:

$$N_2 + O \rightleftarrows NO + N \qquad (1)$$

$$N + O_2 \rightleftarrows NO + O \qquad (2)$$

Reactions 1 and 2 contribute approximately equal amounts of NO ([4]). Given time, these reactions continue to an equilibrium level, which is influenced by variables such as flame temperature, residence time, concentration of each gas, and movement of the gases through zones of different temperatures, pressures, and concentrations. Once NO is formed, the rate of decomposition is slow under ordinary reaction conditions. As a result, NO concentration is frozen in the com-

B. K. Thota is now at DuPont, Charleston, WV.
D. T. Pratt is at University of Michigan, Ann Arbor, MI 48109.

0065-8812-81-4180-0211-$2.00

bustion products after they leave the high temperature zone. The NO thus formed can further react with oxygen to form NO_2 as follows:

$$NO + 1/2\ O_2 \rightleftharpoons NO_2 \tag{3}$$

The above reaction demonstrates the coexistence of NO and NO_2. The stability of NO_2 decreases with increasing temperature (5). Consequently, only a small percentage of the total nitrogen oxide emissions is NO_2.

The rate of NO formation from N_2 and O_2 is very highly temperature-dependent. Zeldovich (6) gives the following expression for the rate of formation of NO, derived on the basis of the chain mechanism:

$$\frac{d\ [NO]}{dt} = 3.0 \times 10^{14} e^{-129,000/RT} [N_2][O_2]^{1/2}$$

where: [NO], $[N_2]$, $[O_2]$ = concentrations, gram mole/cm^3; t = time, seconds; T = temperature, degrees Kelvin; R = gas constant, cal/gram mole-°K.

The high activation energy for NO formation accounts for the extreme dependence of reaction rate on temperature. The gas-phase decomposition has a lower, but still appreciable activation energy.

NO_x formation also varies according to the type of combustion involved. In a homogeneous combustion process, fuel and air are intimately and uniformly mixed before combustion. In the case of the heterogeneous combustion process typical of the diesel and gas turbine engines, in which fuel and air mix and burn simultaneously, the chemical reactions resulting in NO formation are unchanged but the physical environment in which these reactions proceed is quite different from that of the homogeneous system (7).

The theoretical prediction of NO (or NO_x) production in a gas turbine combustor is an exceedingly difficult problem since the reaction rate and, hence, the total NO_x formed depends on the time-temperature history of the reactants (8). The time-temperature history is difficult to determine since the fluid dynamics of the combustor reaction zone is complex and not well understood. For engineering purposes, a simple correlation is often used to predict the effect of various operating variables on NO_x emissions. This correlation is of the form:

$$NO_x = A\ P_4{}^a T_3{}^b \exp (T_3/c)\ f^d$$

where A, a, b, c, d are constants, P_4 is the combustor pressure, T_3 is the combustor air inlet temperature and f is the overall combustor fuel-to-air mass ratio. An additional multiplier needed is τ, the residence time.

Decomposition of NO

The reactions influencing nitric oxide decomposition are the bimolecular atom exchange reactions between nitric oxide and nitrogen atoms, between nitric oxide and nitrogen atoms, between nitric oxide and oxygen atoms (the reverse of the NO formation reactions), and between pairs of NO molecules.

$$NO + N \rightarrow N_2 + O \tag{4}$$

$$NO + O \rightarrow N + O_2 \tag{5}$$

$$NO + NO \rightarrow N_2O + O \tag{6}$$

NO decomposition by Reactions 4 and 5 is about 500 times more than that by Reaction 6. Reaction 5 is endothermic and therefore much less favorable energetically than Reaction 4. The large number of oxygen atoms relative to nitrogen atoms present in the system tends to compensate for this, however, and the Reactions 4 and 5 proceed at about the same forward rate (7).

As in all chemical reactions, the rates of formation and decomposition for NO can be hastened by means of catalysts; while catalysts cannot change equilibrium concentrations, they reduce the time of attainment of equilibrium. A number of investigators have studied the NO decomposition reaction and they have come out with different results (9, 10, 11). Some (9, 12) have indicated that NO decomposition is heterogeneous below 1000°K. They also suggest Reactions 4, 5, and 6 as the only possible reactions, although it is not agreed as to which reaction is predominant. It also appears that there is some variation and uncertainty in the kinetic parameters presented by various investigators (13).

Control Techniques for NO_x Emissions

Some of the methods used for reducing NO_x emissions include limiting fuel nitrogen content, low excess air firing, staged combustion, "off-stoichiometric firing," flue gas recirculation, water and steam injection, selective or nonselective catalytic reduction as well as combinations of these techniques 3, 14 to 19, 46).

Catalytic Treatment of NO_x

Catalytic treatment of exhaust gases to control NO_x emissions is a viable and possible technique. Two major treatment processes are catalytic decomposition of NO and catalytic reduction of NO.

Decomposition of NO to its elements appears to be thermodynamically favorable in the temperature range of 300 to 4000° K (13). The decomposition reaction is

$$2 NO \rightarrow N_2 + O_2 \tag{7}$$

Catalytic reduction of NO is achieved by mixing a reducing agent such as CH_4, H_2, or CO into the gas and passing the mixture over a catalyst in a chamber. Some of the possible chemical reactions are

$$2NO + 2CO \rightarrow N_2 + 2CO_2 \tag{8}$$

$$2NO + 2H_2 \rightarrow N_2 + 2H_2O \tag{9}$$

The reaction chamber effluent contains mostly N_2, CO_2, and H_2O.

More than 600 NO_x catalysts on various supports have been developed and evaluated on automotive exhaust systems. Compounds of 36 individual metals or combinations and several alloys have been used to make catalysts on about 20 different supports. In tests involving automotive exhaust gases at space velocities in the range of 15,000 to 50,000 Hr^{-1}, none was found to be effective for NO decomposition (46). Many of the above catalysts, however, proved to be excellent promoters for the reduction of NO_x by CO, H_2, or hydrocarbons (HC) present in exhaust gases. The life of a catalyst is limited by the temperature to which it is exposed, the oxidizing tendency of the flue gas, attrition losses during operation, and presence of poisons.

In addition to observing catalytic NO removal when a CO-NO-N_2 mixture was reacted, Taylor (20) also found extensive NO removal (up to 77%) from NO-N_2 mixtures at a temperature of 215° C without any added reducing agent. The catalyst employed was alumina-supported barium-promoted copper chromite. Sakaida, et al. (21), using a platinum-nickel-alumina catalyst determined that NO decomposition was second order at 425° to 540°C and 1 to 15 atmospheres. Sourirajan and Blumenthal (22) studied NO_x reduction by CO and H_2 between 100° to 550° C at compositions of 300 to 1500

ppm and reported CuO-silica to be the best catalyst. Roth and Doerr (23) continued on the work of Taylor using supported CuO, CuO-Cr_2O_3 catalysts. Baker and Doerr (4) reported the formation of NH_3 in the presence of water vapor. Ayen and Peters (24) investigated NO reduction with hydrogen. In their work, Jones et al. (25), showed that some selectivity toward NO reduction by H_2 on noble metals is obtained at stoichiometric engine operations. Bauerle, et al. (26), studied the reduction of NO by CO in the presence of O_2 at 320°C using several copper-based and noble metal catalysts.

In another study using infrared spectroscopy, London and Bell (27) studied the reduction of NO by CO on silica supported copper oxide and postulated that nitric oxide can dissociate upon adsorption. Amirnazmi, et al. (28), studied the aspect of oxygen inhibition on the rate of NO decomposition on metal oxides and platinum between 450° and 1000°C, around atmospheric pressure. They found that the reaction was first order with respect to NO and that above 450°C, N_2 and O_2 were the only products. Wise and French (29) reported NO decomposition to be second order in the range of 600° to 1000°C, the reaction being surface-catalyzed below 730°C and homogeneous at higher temperatures. Yuan, et al. (12), have shown NO removal to be heterogeneous in a zero-order reaction below 1100°C, and homogeneous in a second order reaction above 1400°C. Between 1100°C and 1400°C, both heterogeneous and homogeneous reactions took place simultaneously, the contribution from each depending on the area of the surface, and its chemical nature, as well as on the temperature and concentration. In their study of flash desorption of nitric oxide from tungsten, Yates and Madey (30) found that at high temperatures, NO was dissociated on the surface with an activation energy of 46.7 K-cal $mole^{-1}$.

Klimisch and Taylor (31) noticed dual functionality and synergistic effects with catalysts containing nickel in combination with copper, platinum, or palladium. Among these catalysts the ammonia decomposition function resides primarily in the nickel while the reduction activity is obtained from the other metal. Bartholomew (32) reported that monolithic-supported Pd-Ni and Pd-Ru catalysts are effective in NO_x removal; but he indicated that longterm stability of these catalysts is a serious problem. CO and C_3H_6, along with the NO_x, were used in the gas streams and the temperature range in these investigations was 480°C to 600°C. Recently, General Motors Research Laboratories (33) have come up with a report on catalytic NO reduction studies and mentioned that ruthenium selectively catalyzes NO conversion reaction, practically without any ammonia formation. This work was performed in a reducing atmosphere using both hydrogen and carbon monoxide. The maximum temperature employed in this case was around 600°C. Several catalysts have been offered for NO_x control depending on NO_x sources, concentrations, conditions, etc., and most of these are covered by patents (34). In Table 1 is given a summary of catalytic NO studies by various authors.

In spite of these investigations, there seem to be wide areas of disagreement between the published results presented by different authors. In addition, not much significant work has been done in the area of catalytic studies relevant to NO_x control from turbine emissions, which typically occur at temperatures around 1000°C and higher and roughly at atmospheric pressure.

EQUIPMENT AND PROCEDURE

Flow Equipment

In the flow arrangement shown in Figure 1, gases first entered the mixing chamber, then the preheater. From the preheater, the gas mixture was directed into the reactor. The equipment was designed such that continuous analysis of the inlet and exit gas streams for NO_x was possible.

The mixing chamber was a 30 cm long, 3.4 cm I.D., and 0.64 cm thick lucite tube. A ceramic tube, 2.54 cm I.D. and 54 cm long, containing 0.64 cm Intalox-saddle packing served as the gas preheater. The preheater tube was placed in a horizontal fashion in a Lindberg Hevi-Duty Type 167 electric furnace, which was capable of heating up to 1000°C.

The reactor section consisted of a 2.7 cm I.D., 76 cm long Leco ceramic tube with one end tapered (Figure 2). The actual catalytic reactor was a 10 cm cylindrical cage of diameter very close to the i.d. of the reactor tube and made of #16 stainless steel screen (16 mesh). The cage was supported on a tube made up of a threaded portion 14 cm long 0.95 cm O.D. stainless steel tube, and a 30 cm long 0.64 cm O.D. stainless steel tube, joined together by plasma welding. The end of the support tube near the catalyst cage was sealed. The cage was held in position onto the support tube by means of stainless steel nuts. In

subsequent runs, a slight modification of this arrangement was made. The catalyst cage was replaced by just two end screens to hold the catalyst in place. A chromel-alumel thermocouple insulated by a series of 5.0 cm long double-hole ceramic insulators, was embedded in the catalyst bed. The thermocouple wires with insulators were run through the support tube. The chromel-alumel thermocouple was stipulated by the manufacturer (Hoskins Mfg. Co.) to be accurate to within ±3/4 percent in the temperature range of 280 to 1269°C. The open end of the support tube was sealed using Sauereisen electro temp cement No. 8, after the thermocouple was connected to a D.C. millivoltmeter.

The inlet sample was collected using a 0.32 cm O.D. stainless steel sample probe installed just at the upstream (inlet) end of the catalyst bed. The inlet sample probe, the support tube, and a 0.48 cm O.D. stainless steel tube, used as exhaust outlet, all were incorporated in the stainless steel seal assembly.

The tapered end of the reactor tube was connected to the gas line from preheater, by means of a Cajon ultra-torr adaptor and Swagelok connections. The sealing of this end was ensured using Sauereisen cement.

The reactor tube was heated by means of a Lindberg CF-1 combustion tube laboratory furnace. The voltage on the heating elements could be varied by means of built-in transformers. A heating length of about 20 cm was obtainable in this furnace. The reactor tube was horizontally suspended in the furnace so that it was coaxial with the heating section of the furnace. This arrangement enabled the reactor tube and hence the catalyst bed to be exposed to uniform heating.

The exhaust line from the reactor was led into a hood vent by means of a 0.48 cm O.D. stainless steel tube. This exhaust line was branched off at the reactor unit and led to the analyzing equipment for monitoring exit gases. The pressure of the gas stream entering the reactor tube was measured by a pressure gauge. A platinum-platinum +13% rhodium thermocouple measured the temperature of the reactor tube.

Gas Analyses

The inlet and the exit sample lines from the reactor tube were led through the Thermo-electron Model 44 Chemiluminescent $NO-NO_x$ analyzer. Since the NO_x analyzer measured only NO_x concentrations in the gas stream, the other components were analyzed from grab samples using a Carle Model 8000 gas chromatograph with a thermal conductivity detector. The chromatograph was equipped with two columns--a Molecular Sieve 5A column of stainless steel 300 cm long and 0.64 cm O.D. and a Porapak S column again of stainless steel, 180 cm long, 0.64 cm O.D. The Molecular Sieve 5A separates N_2, O_2 and CO from other gaseous components while the Porapak S column separates CO_2 and N_2O. The flow rate of helium carrier gas was maintained at 16 ml/min through the Molecular Sieve column and 25 ml/min through the Porapak S column while the oven temperature was maintained at 62°C.

Catalyst Preparation and Analyses

Supported nickel and cobalt catalysts were used in the experimentation. Both these catalysts were prepared by impregnation. Nickel and cobalt were deposited in oxide form from their salts $Ni(NO_3)_2 \cdot 6H_2O$ (Mallinckrodt analytical reagent) and $Co(NO_3)_2 \cdot 6H_2O$ ("Baker Analyzed" reagent) on -4 to 10 mesh (U. S. standard sieve) GGB grade zirconia chips supplied by the Zircoa Company. After impregnation the raw catalysts were calcined in nickel crucibles over a Bunsen burner and kept in an oven overnight to drive away any residual moisture. These were subsequently heated in an electric furnace at about 1000°C for approximately 90 minutes. This calcination step converted the deposited material to the desired formulation.

The catalysts were analyzed for the active component composition according to the methods outlined in Furman (45); dimethyl glyoxime precipitation for nickel and a α-nitroso-β-naphthol precipitation for cobalt. The composition of the nickel catalyst was found to be 3.74% by weight of nickel, while that of the cobalt catalyst was 3.62% by weight of cobalt. It was felt useful to analyze the zirconia support material also in order to subsequently investigate the catalytic activity of the support material itself. The analysis through mandelic acid precipitation, described in Furman (45), was followed and this yielded 91.8% by weight of zirconia (ZrO_2) in the support, the rest of the compounds being SiO_2, CaO, etc. The manufacturer of the support material (Zircoa Corporation) reports the composition to be 93.5% by weight of ZrO_2.

Analysis was also performed on the used nickel catalyst and this revealed that the nickel content had decreased by a small amount

(close to 5%).

When steady conditions were established with N_2 flow, test runs were carried out with the nickel catalyst, using $NO-N_2$ gas mixtures and occasionally $NO-N_2$-He mixtures, in the absence of any reducing agent (Runs 1 to 16). This was done to investigate the possibility of thermal decomposition of NO catalyzed by the surface, and also to determine the role of the catalyst in effecting the decomposition reaction. Seventy grams of catalyst were used in the bed and the volume of the bed was calculated to be 44 cc, which remained approximately constant in the course of experimentation. This volume remained constant with the cobalt catalyst also.

After steady state was reached for each run (approximately 10 minutes were allowed), the inlet and exit gases were analyzed for NO_x concentrations. As a cross check, one set of samples (consisting of inlet and exit gases) was collected in two plastic bags and analyzed in Beckman Chemiluminescent NO/NO_x Analyzer. The results obtained with this instrument were quite close (95%) to those obtained with the Thermoelectron Chemiluminescent Analyzer. The chemiluminescent analyzer was frequently checked for its calibration with standard $NO-N_2$ calibration gas, during the course of the work.

In some subsequent runs (Runs 20 to 24), CO was added to the gas streams and changes in concentration of NO_x in inlet and exit samples were monitored. Additionally, two sets of grab samples were collected and analyzed in the gas chromatograph.

For the cobalt catalyst runs, 67 gms of the catalyst were used. $NO-N_2$-He mixtures (Runs 30 to 36) and also $NO-N_2$-CO (Runs 40 to 57) were used as feed streams. A few runs were also carried out to observe the effect of the presence of oxygen on NO decomposition (Runs 60 to 64).

A few runs were carried out to investigate the catalytic effect of the zirconia support alone (Runs 80 to 86) and also to observe the possibility of an uncatalyzed decomposition reaction of NO (Runs 70 to 74). Approximately 80 grams of zirconia chips (-4 + 10 mesh U.S. standard sieve) were used in the experimentation with zirconia. The volume of the catalyst bed was 50 cc. At steady state, using $NO-N_2$-He feed mixture, inlet and exit NO_x compositions were monitored. Next, the zirconia was removed from the reactor tube and blank runs were carried out using $NO-N_2-He$ mixture. Again the inlet and exit gases were monitored through the chemiluminescent analyzer and readings taken at steady state.

RESULTS AND DISCUSSION

The variables studied in the current investigation were the inlet NO_x concentration, the gas hourly space velocity (GHSV) and the effect of the presence of oxygen in $NO-N$ gas mixture was also investigated, although only limited data were obtained in this case. The inlet NO_x concentration was generally varied from 100 to 1700 ppm, although occasionally higher NO_x concentrations were used. Gas hourly space velocity in the range of 8,000 to 35,000 hr^{-1} was used. Gas hourly space velocity is defined here as volume of gas per hour at 20°C and 1 atmosphere pressure per volume of catalyst bed.

Performance of Supported Nickel Catalyst

The effect of variation of inlet NO_x concentration at a constant space velocity and also variation of space velocity at a constant inlet NO_x concentration on NO_x decomposition is shown in Figures 3 and 4 (Runs 1 to 9, 15 from Table 2). The decrease in NO_x removal with increasing space velocity may be due to a low residence time. By assuming one or both of the following reactions to be occurring,

$$2NO \rightleftharpoons N_2 + O_2$$

$$2NO \rightleftharpoons N_2O + O$$

the prevailing chemical reaction(s) appears to have reached equilibrium under given conditions with maximum NO_x conversion being closer to 70%. Some authors (41, 29, 12) have confirmed that the above reactions are the only possible reactions in the temperature ranges encountered in the present investigation. In addition, Amirnazmi, et al. (28), have reported that at high temperatures (>450°C) N_2 and O_2 are the only products of NO decomposition.

When CO is introduced into the system in excess of the stoichiometric quantities (for reacting with NO), nearly complete reduction of NO occurs as is evident from Table 3 and Figure 5. The NO_x removal seems to be nearly independent of both inlet NO_x concentration and space velocity, although at lower inlet concentrations, there is a slight decrease in NO_x reduction. The results of the chromato-graphic analysis of the two sets of grab samples are given in Table 4. Two inferences can be drawn from these results: (a) The analysis clearly shows that a considerable amount of CO_2 is present in the inlet gases even before the gases enter the reaction section. This implies that appreciable reduction of NO with CO is occurring prior to its entry into the catalyst bed. The actual inlet concentrations (at the mixing tube end) of NO_x, CO, etc., are different from those obtained in sampling. (b) The chromatographic analysis is not very accurate for this type of work as the concentrations of the components of interest are always in the ppm range. It is also suspected that the grab samples were contaminated with atmospheric air--as is evident from the presence of oxygen in the samples. Thus the analyses results obtained are more of the qualitative type. The following chemical reaction is probably taking place as the literature shows (22).

$$2CO + 2NO \xrightarrow[cat.]{} 2CO_2 + N_2$$

No detectable N_2O and NO_2 were present in the sample gases. This is to be expected as the concentration of NO_x in the exit gases is very low because of high conversion.

The above reaction is thermodynamically favorable even at temperatures as high as 1300°K (22) although scarcely any work has been done hitherto in this direction. Concentrations of CO in far excess of those required by the stoichiometry have been used in the current investigation. This was because pure CO from the cylinder was used as a source of reducing agent and further reduction in its concentration was not possible. Presence of excess CO is not unrealistic of the actual gas turbine exhausts where the variation of CO emissions (in relation to NO_x) can be anywhere between 1 and 50 times that of NO_x emissions (42, 43).

By chemical reduction with CO, not only most of the NO_x is removed but part of the CO itself is oxidized to CO_2. This is desirable as both NO_x and CO are toxic and harmful emissions.

Another possible reaction indicated by Kirk and Othmer (44) is the oxidation of CO by the oxides of cobalt, copper, nickel, etc. in the temperature range of 300 to 1500°C according to the following reaction

$$CO + MO \rightarrow CO_2 + M$$

where MO is the metal oxide and M is the re- duced form of the metal oxide. Roth and Doerr (23) have indicated that copper chromite and copper oxide catalysts were capable of existing in both oxidizing and reducing states, although their investigation was confined to lower tem- peratures. Whether the above-mentioned reac- tion is taking place or not can be determined by analyzing the catalyst for its initial and final oxygen content or by accurate monitoring of CO and CO_2 concentrations by passing CO over the catalyst. The same authors have also mentioned that carbon monoxide is capable of undergoing decomposition according to the fol- lowing reaction

$$2CO \rightleftharpoons C + CO_2$$

Carbon monoxide under a pressure of 10 to 40 atmospheres and a temperature of 700 to 1040°C has been found to produce considerable carbon which has an autocatalytic effect. Even though equilibrium for decomposition is favorable, the reaction rate may be slow. In fact, above 800°C, the decomposition is not favorable. It is possible that part of the CO_2 formed in the process of the current in- vestigation is coming from the decomposition of CO. The occurrence of both of the above reactions could be the reason for the presence of an excess amount of CO_2 over that expected from the stoichiometry, as is found in the analytical results. However, the pressure employed in this work is far less than the above mentioned range and chances of the above reaction taking place are consequently less.

Youn (37) has also studied NO_x conversion by CO, using catalysts such as 0.5% Pd on alumina and 3.2% Cu-1.8% Ni on alumina. His GHSV was 30,000 and the inlet gas also con- tained 0.5% O_2. Good conversions were ob- tained only when the CO concentration exceeded the stoichiometric requirement for the reac- tion with both O_2 and NO_x. When CO concentra- tion was just high enough for the reaction only with NO_x, a maximum of only 50% NO_x con- version was observed. The conversions were favorable in the temperature range 375 to 475°C, while below this range they were poor.

Performance of Supported Cobalt Catalyst

When no CO was present in the reaction gas stream, the NO_x removal was very low-- usually not exceeding 15% (Table 5, Figure 6). The conversion is 3 to 4 times less than that obtainable with the nickel catalyst. This may be due to the fact that different catalysts have different catalytic activities and that selectivity under given conditons is a domi- nant and decisive factor for a catalyzed chem- ical reaction. Furthermore, the catalyst characteristics and the mechanism of NO de- composition on cobalt catalyst may be dif- ferent from that on the nickel catalyst, al- though both the catalysts were prepared under nearly similar conditions.

When CO was added to the inlet gas streams, results similar to that with nickel catalyst were obtained. A phenomenal in- crease in NO_x removal is evidenced (Table 6, Figure 7). As in the case of the nickel catalyst, NO_x removal with CO is seen to be nearly independent of both inlet NO_x concen- tration and space velocity. Results of gas chromatographic analysis of the samples are shown in Table 7. Oxides of nitrogen other than NO_x (NO_x measured with chemilumines- cent analyzer) could not be detected with the chromatograph. It is assumed that these are not present.

When oxygen was introduced into the NO-N_2 reaction mixture, in general a reduction in the decomposition of NO was observed (Table 8). Amirnazmi, et al. (28), also observed the oxygen inhibition effect on the decom- position of NO. It was possible to analyze only one set of samples in the gas chromato- graph and the output from the chromtograph indicated the presence of N_2 and O_2.

Uncatalyzed Thermal Decomposition of NO_x

Some thermal decomposition in the ab- sence of any catalyst seems to be taking place (Table 9, Figure 8), although this is not high. Variation of inlet NO_x concentra- tion does not seem to have much effect on NO_x removal. Some authors (13, 9, 12) have mentioned that in the temperature range of 1000 to 1400°C, the NO decomposition reaction is a combination of heterogeneous and homo- geneous reactions, whereas above 1400°C the reaction is entirely a homogeneous gas phase reaction. The observations made in this study seem to agree to a limited extent with the above theory. It is possible that part of the decomposition reaction is being cata- lyzed by certain sections of the reactor such as the tube walls, support tube, etc.

Catalytic Activity of Zirconia Support

Data and results obtained with plain zirconia chips as the catalyst bed are given in Table 10 and Figure 9. The results clearly show that zirconia by itself has some catalytic effect. It is also seen that here too the NO_x removal decreases with increasing space velocities.

It is interesting to note here that catalytic decomposition of NO_x on zirconia-supported cobalt catalyst is, in most cases, less than the uncatalyzed (blank experiment) and also zirconia catalyzed reactions. While an explanation for the discrepancy in the results of uncatalyzed conversions is lacking, the difference in NO_x removal by zirconia supported cobalt catalyst as compared to the plain zirconia may be due to the reasoning that the catalytic activity of the combined form--namely cobalt oxide supported on zirconia, is different from the individual activities of the two constituents. It appears that the supported cobalt catalyst is retarding the decomposition reaction, when seen in relation to uncatalyzed and zirconia-catalyzed reactions. Probably the catalytic activity of the zirconia itself is retarding the overall activity of the supported cobalt catalyst.

CONCLUSIONS

The following conclusions can be drawn from the current investigation:

1. Significant results in the NO_x reduction were obtained on both zirconia supported nickel and zirconia supported cobalt catalysts, at temperatures in the vicinity of 1000°C.

2. Around 1000°C, the decomposition of NO_x seems to be the combined effect of heterogeneous and homogeneous reactions.

3. Zirconia supported nickel catalyst is more effective for NO_x decomposition than similarly supported cobalt catalyst.

4. When carbon monoxide is introduced in excess into the reaction streams, in the presence of both supported nickel and supported cobalt catalysts, remarkable reduction of NO_x (close to 100%) is achieved. This reduction is found to be independent of the space velocity and the inlet NO_x concentration.

5. Limited data obtained on NO_x decomposition on a supported cobalt catalyst in the presence of oxygen, indicate that oxygen has a retarding effect on NO_x decomposition.

LITURATURE CITED

1. Hopper, T. G. and W. A. Marrone, "Impact of new source performance standards on 1985 national emissions from stationary sources," EPA 450/3-76-017 (May 1976).

2. McCutchen, G. D., Chem. Eng. Prog., 73(8), 58 (1977).

3. Cichanowicz, J. E., EPRI Journal, 4(4), 22 (1979).

4. Baker, R. A. and R. C. Doerr, Ind. Eng. Chem., Process Des. Develop., 4, 188, (1965).

5. U. S. Department of Health, Education, and Welfare, "Control techniques for nitrogen oxide emissions from stationary sources," NAPCA Publication No. AP-67 (1970).

6. Zeldovich, J., Acta Physicochim., 21, 577 (1946).

7. Springer, G. S. and D. J. Patterson (Eds.), Engine Emissions, Plenum, New York (1973).

8. Sullivan, D. A. and P. A. Mas, "A Critical review of NO_x correlations for gas turbine combustors," Paper #75-WA/GT-7, ASME Winter Annual Meeting, Houston (Nov.-Dec., 1975).

9. Fraser, J. M. and F. Daniels, J. Phy. Chem., 62, 215 (1958).

10. Freedman, E. and J. W. Daiber, J. Chem. Phys., 34, 1271 (1961).

11. Glick, H. S., J. J. Klein and W. Squire, J. Chem. Phy., 27, 850 (1957).

12. Yuan, E. L., J. I. Slaughter, W. E. Koerner and F. Daniels, J. Phy. Chem., 63, 952 (1959).

13. Baulch, D. L., D. D. Drysdale, D. G. Horne and A. C. Lloyd, Evaluated Kinetic Data for High Temperature Reactions, Vol. 2, Butterworths, England (1973).

14. Hunter, S. C. and W. A. Carter, Chem. Eng. Prog., 73(8), 66 (1977).

15. Siddiqui, A. A., J. W. Tenini and L. D. Killion, Hydrocarbon Processing, 55(10), 94 (1976).

16. Anon., Chem. Eng., 85, 85 (June 19, 1978).

17. Anon., Chem. Eng., 85, 23 (July 3, 1978).

18. Anon., *Chem. Eng.*, 84, 70 (April 25, 1977).

19. Johnson, R. H. and C. Wilkes, "Environmental considerations - 1975," Publication #GER-2486D, General Electric Co., Schenectady, N. Y. (1976).

20. Taylor, F. R., "Elimination of oxides of nitrogen from automobile exhaust," Air Pollution Foundation Report No. 28, San Marino, CA (1959).

21. Sakaida, R. R., R. G. Rinker, U. L. Wang and W. H. Corcoran, *AIChE Journal*, 7, 658 (1961).

22. Sourirajan, S. and J. L. Blumenthal, *Int. J. Air Wat. Poll.*, 5, 24 (1961).

23. Roth, J. and R. Doerr, *Ind. Eng. Chem.*, 53, 293 (1961).

24. Ayen, R. J. and M. S. Peters, *Ind. Eng. Chem., Process Des. Develop.*, 1, 204 (1962).

25. Jones, J. H., J. T. Kummer, K. Otto, K. Shelef and E. E. Weaver, *Env. Sci. Tech.*, 5, 790 (1971).

26. Bauerle, G. L. G. R. Service and K. Nobe, *Ind. Eng. Chem., Prod. Res. Develop.*, 11, 54 (1972).

27. London, J. W. and A. T. Bell, "A Simultaneous infrared and kinetic study of the reduction of nitric oxide by carbon monoxide over copper oxide," Paper submitted to *J. Catalysis*, (March 1973).

28. Amirnazmi, A., J. E. Benson and M. Boudart, "Oxygen inhibition on the decomposition of NO on metal oxides and platinum," Paper submitted to *J. Catalysis* (1973).

29. Wise, H. and M. F. Frech, *J. Chem. Phy.*, 20, 22, 22, 1724 (1952).

30. Yates, J. T. and T. E. Madey, *J. Chem. Phy.*, 45, 1623 (1966).

31. Klimisch, R. L. and K. C. Taylor, "Dual functionality and synergism in the catalytic reduction of nitric oxide," Research Publications of General Motors, Publication No. GMR-1195 (1972).

32. Bartholomew, C. H., *Ind. Eng. Chem., Prod. Res. Develop.*, 14, 29 (1975).

33. Klimisch, R. L. and K. C Taylor, *Ind. Eng. Chem., Prod. Res. Develop.*, 14, 26 (1975).

34. Lawrence, A. A. (Ed.), "Nitrogen oxides emission control," Noyes Data Corp., New Jersey (1972).

35. Bauerle, G. L. and K. Nobe, *Ind. Eng. Chem., Prod. Res. Develop.* 13, 185 (1974).

36. Bauerle, G. L., L. L. C. Sorensen and L. Nobe, *Ibid*, 13, 61 (1974).

37. Youn, K. C., *Hydrocarbon Processing*, 58(2), 117 (1979).

38. Lamb, A. and E. L. Tollefson, *Can. J. Chem. Engr.*, 53, 68 (1975).

39. Thomas, T. R. and D. T. Pence, "Reduction of NO_x with ammonia over zeolite catalysts," Report to USAEC, contract #AT(10-1)-1375 S-72-1 (1974).

40. Yamaguchi, M., K. Matsuhita and K. Takami, *Hydrocarbon Processing*, 55(8), 101 (1976).

41. Kaufman, F. and J. R. Kelso, *J. Chem. Phy.*, 23, 1702 (1955).

42. Sawyer, R. F. and E. S. Starkman, "Gas turbine exhaust emissions," Selected Papers on Vehicle Emissions Part III, Paper No. 680462, Soc. Auto. Engineers, 631 (1967-70).

43. Smith, D. S., R. F. Sawyer and E. S. Starkman, *J. Air Poll. Contr. Assoc.*, 18, 30 (1968).

44. Kirk, R. E. and Othmer, D. F. (Eds.), *Encyclopedia of Chemical Technology*, 2nd edition, Vol. 4, Interscience, New York (1964).

45. Furman, N. H. (Ed.), *Standard Methods of Chemical Analysis*, Vol. 1, 6th edition, Van Nostrand, New York (1964).

46. Hall, H. and W. Bartok, *Env. Sci. Tech.*, 5, 320 (1971).

Table 1. Summary of Some Previous Catalytic NO Conversion Studies

Investigator(s)	Catalyst	Reactor Type	Temperature Range, °C	Pressure Range psia	Gas Mixture	Reaction Order With Respect to NO	Reference
Taylor	Barium-promoted copper chromite	Flow	215	-	$NO-N_2$	-	20
Bauerle and Nobe	Mixed metal oxides	Flow	25-520	-	$NO-H_2-N_2$	-	35
Fraser and Daniels	Metal oxides	Flow	740-1040	14.7	$NO-He$	0	9
Sakaida et al.	Supported Pt-Ni	Flow	425-540	14.7-29.7	$NO-N_2$	2	21
Sourirajan and Blumenthal	CuO-Silica	Flow	250-1000	-	$NO-CO$	-	22
Roth and Doerr	Supported Cr_2O_3, CuO	Flow	104-249	-	$CO-NO-N_2$	-	23
Baker and Doerr	$CuO-Cr_2O_3$	Flow	115-270	-	$NO-CO-N_2$	-	4
Ayen and Peters	Copper-zinc-chromia	Flow (differential)	375-425	-	$NO-N_2-H_2$	-	24
ıan et al.	Alundum reactor	Flow	700-1800	14.7	$NO-N_2-He$	2 (above 1400°C); 0 (below 1100°C)	12
ıuerle et al.	Copper-based, noble metals	Flow	150-400	-	$NO-CO$	-	26

Table 1. Summary of Some Previous Catalytic NO Conversion Studies (Continued)

Investigator(s)	Catalyst	Reactor Type	Temperature Range, °C	Pressure Range psia	Gas Mixture	Reaction Order With Respect to NO	Reference
Bauerle et al.	Supported Cu-Ni	Flow	100-500	-	NO-CO	Variable	36
London and Bell	Copper oxide on silica	Recycle	135-200		NO-CO-He	-	27
Amirnazmi et al.	Metal oxides, Pt	Flow	450-1000	14.7	$NO-O_2$-He	1	28
Wise and French	Quartz reactor	Static	600-1000	-	$NO-O_2$	2 (up to 1000°C)	29
Klimisch and Taylor	Ruthenium	Flow	200-600	-	$NO-CO-H_2-CO_2-N_2$	-	33
Bartholomew	Supported Pd-Ni, Pd-Ru	Flow	480-600	-	$NO-C_3H_6-O_2-H_2O-N_2$	-	32
Youn	Pt on γ-alumina	Flow, with Reduction/Oxidation	100-450	-	(1)$NO-NH_3$ (2)NO-CO (3)$NO-CH_4$	-	37
Lamb and Tollefson	Brass, Monel	Flow	340-560	14.7	$NO-H_2-He-O_2$	-	38
Thomas and Pence	Zeolite	Flow	350	-	$SO_2-NO-CO_2-H_2O-NH_3$	-	39
Yamaguchi, et al.	Rare Earth on Al_2O_3	Flow	480	100	NH_3-NO	-	40

TABLE 2

NO$_x$ removal from NO-N$_2$-He mixtures using supported nickel catalyst

Run #	NO$_x$ Concentration (ppm)		Percent NO$_x$ Removal	Reactor Temperature (°C)	Preheater Temperature (°C)	Flow Rate of Gas Mixture at 1 atm, 20°C (m^3/hr)	Gas Hourly Space Velocity (Hr^{-1})
	Inlet	Exit					
1	1,550	550	64.5	1160	480	0.527	12,000
2	1,550	990	36.1	1160	510	0.903	20,400
3	1,500	900	40.0	1160	540	0.875	19,800
4	1,500	810	46.0	1160	540	0.716	16,200
5	1,500	800	46.7	1160	540	0.830	18,800
6	1,130	650	42.5	1120	480	0.932	21,100
7	1,070	670	37.4	1110	430	0.937	21,200
8	1,000	600	40.0	1110	450	0.926	21,000
9	890	525	41.0	1110	450	0.923	20,900
10	790	460	41.8	1110	430	1.107	17,300
11	675	395	41.5	1110	450	0.974	22,100
12	560	225	59.8	1110	510	0.660	15,100
13	480	255	46.9	1160	480	0.850	19,300
14	390	218	44.1	1110	450	0.966	21,900
15	258	158	38.8	1110	430	1.039	23,600
16	200	127	36.5	1110	430	1.107	25,100

TABLE **3**

Effect of the presence of CO on NO$_x$ removal on supported
nickel catalyst

Run #	NO$_x$ Concentration (ppm)		Percent NO$_x$ Removal	Reactor Temperature (°C)	Preheater Temperature (°C)	Flow Rate of Gas Mixture at 1 atm, 20°C (m³/hr)	Gas Hourly Space Velocity (Hr⁻¹)
	Inlet	Exit					
20	940	6.0	99.4	1080	480	0.725	16,400
21	590	5.8	99.0	1080	450	0.736	16.600
22	770	6.0	99.2	1080	480	0.850	19,200
23	500	6.0	98.8	1070	480	0.838	18,900
24	125	6.2	95.0	1070	480	0.810	18,300

TABLE **4**

Gas analysis of the samples from tests with supported nickel catalyst using
NO-CO-N$_2$ feed streams (test conditions included)

Run #	Inlet/ Exit	Reactor Temperature (°C)	GHSV (Hr⁻¹)	Gas Composition					% NO$_x$ Removal
				NO$_x$ ppm	CO ppm	CO$_2$ ppm	O$_2$ Mole%	N$_2$ Mole%	
20	Inlet			940	5,130	626	8.00	91.00	
		1080	16,400						99.4
	Exit			6	3,500	2,270	8.75	90.40	
21	Inlet			590	8,900	1,640	2.70	95.5	
		1080	16,600						99.0
	Exit			5.8	5,800	3,440	4.00	95.5	

TABLE 5

NO_x removal from $NO-N_2-He$ mixtures using supported cobalt catalyst

Run #	NO_x Concentration (ppm)		Percent NO_x Removal	Reactor Temperature (°C)	Preheater Temperature (°C)	Flow Rate of Gas Mixture at 1 atm, 20°C (m^3/hr)	Gas Hourly Space Velocity (Hr^{-1})
	Inlet	Exit					
30	1,750	1,500	14.3	1070	480	0.912	20,600
31	900	790	12.2	1070	540	0.872	19,800
32	775	700	9.7	1070	450	1.144	25,800
33	750	700	6.7	1040	610	0.671	15,200
34	520	480	7.7	1070	470	1.274	28,800
35	390	365	6.4	1040	590	0.651	14,700
36	283	265	6.4	1040	510	0.997	22,600

TABLE 6

Effect of the presence of CO on NO_x removal on supported cobalt catalyst

Run #	NO_x Concentration (ppm)		Percent NO_x Removal	Reactor Temperature (°C)	Preheater Temperature (°C)	Flow Rate of Gas Mixture at 1 atm, 20°C (m^3/hr)	Gas Hourly Space Velocity (Hr^{-1})
	Inlet	Exit					
40	7,500	1,450	80.6	1070	540	0.606	13,800
41	2,900	60	97.9	1070	560	0.583	13,200
42	1,690	30	98.2	1040	480	1.107	25,100
43	1,660	7.8	99.5	1070	500	0.901	20,300
44	1,425	28	98.0	1040	510	1.102	24,900
45	1,400	6.4	99.5	1070	540	0.736	16,600
46	1,250	7.2	99.4	1070	520	1.082	24,500
47	970	7.3	99.2	1070	520	1.209	27,400
48	660	7.5	98.9	1070	480	1.572	35,500
49	605	6.3	99.0	1070	540	0.617	14,000
50	530	6.8	98.7	1070	540	0.767	17,400
51	475	7.4	98.4	1070	520	1.365	30,800
52	390	6.2	98.4	1050	540	0.753	17,100
53	385	7.3	98.1	1050	480	1.538	34,700
54	268	4.8	98.2	1040	540	1.433	32,400
55	190	4.7	97.5	1040	540	1.289	29,100
56	142	7.0	95.1	1070	480	1.526	34,500
57	100	5.0	95.0	1040	540	1.473	33,300

TABLE 7

Gas analysis of the samples from tests with supported cobalt catalyst
using $NO-CO-N_2$ feed streams

Run #	Inlet/ Exit	Reactor Temperature (°F)	GHSV (Hr-1)	Composition					% NO_x Removal
				NO_x ppm	CO Mole%	CO_2 ppm	O_2 Mole%	N_2 Mole%	
50	Inlet			530	2.80	1,440	0.46	96.10	
		1070	17,400						98.7
	Exit			6.8	2.53	2,350	1.27	95.50	
52	Inlet			390	1.25	978	2.31	96.0	
		1050	17,100						98.4
	Exit			6.2	1.13	1,400	4.38	94.0	
45	Inlet			1,400	1.08	1,300	1.85	98.6	
		1070	16,600						99.5
	Exit			6.4	0.90	3,105	0.90	97.7	

TABLE 8

Effect of the presence of O_2 on NO_x removal on supported cobalt catalyst,
with $NO-N_2-O_2$ gas streams

Run #	NO_x Concentration (ppm)		Percent NO_x Removal	Reactor Temperature (°C)	Preheater Temperature (°C)	Flow Rate of Gas Mixture at 1 atm, 20°C (m^3/hr)	Gas Hourly Space Velocity (Hr-1)
	Inlet	Exit					
60	670	660	1.5	980	590	0.377	8,600
61	525	520	0.9	980	540	1.020	23,100
62	360	348	3.3	980	590	0.818	18,500
63	250	245	2.0	980	570	1.000	22,600
64	195	192	1.5	980	540	1.385	31,400

TABLE   9

Thermal decomposition of NO in the absence of the catalyst (blank runs)
using NO-N$_2$-He feed streams

Run #	NO$_x$ Concentration (ppm)		Percent NO$_x$ Removal	Reactor Temperature (°C)	Preheater Temperature (°C)	Flow Rate of Gas Mixture at 1 atm, 20°C (m^3/hr)	Gas Velocity at 1 atm, (20°C) (m/hr)
	Inlet	Exit					
70	1,310	1,065	18.7	980	480	0.793	1,570
71	1,090	845	22.5	980	540	0.966	1,910
72	800	685	14.4	980	510	1.317	2,600
73	475	410	13.7	980	500	1.263	2,500
74	152	128	15.8	980	480	1.189	2,350

TABLE   10

Catalytic effect of zirconia on the decomposition of NO,
with NO-N$_2$-He feed streams

Run #	NO$_x$ Concentration (ppm)		Percent NO$_x$ Removal	Reactor Temperature (°C)	Preheater Temperature (°C)	Flow Rate of Gas Mixture at 1 atm, 20°C (m^3/hr)	Gas Hourly Space Velocity (Hr^{-1})
	Inlet	Exit					
80	1,225	1,005	18.0	1120	540	0.841	16,700
81	1,040	795	23.6	1070	540	0.779	15,500
82	940	800	14.9	1110	510	0.895	17,700
83	865	750	13.3	1070	510	0.952	18,900
84	690	595	13.8	1070	480	1.011	20,000
85	550	435	20.9	1070	500	0.785	15,600
86	300	225	25.0	1070	500	0.785	15,600

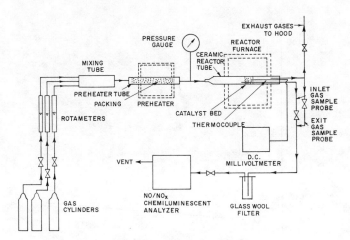

FIGURE I.   FLOW DIAGRAM OF EXPERIMENTAL SET UP

1. CERAMIC REACTOR TUBE
2. CATALYST BED
3. THERMOCOUPLE
4. INLET GAS SAMPLE PROBE
5. SUPPORT TUBE
6. EXHAUST LINE (ALSO EXIT SAMPLE PROBE)
7. STAINLESS STEEL COMPRESSION SEAL

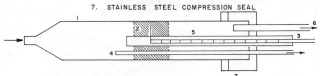

FIGURE 2.   DIAGRAM OF THE CATALYST BED REACTOR

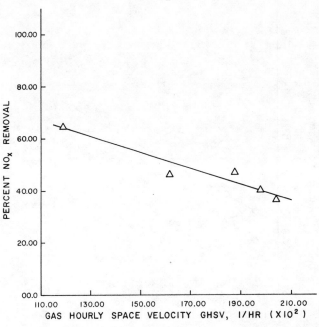

FIGURE 4. EFFECT OF SPACE VELOCITY ON NO_x REMOVAL AT CONSTANT INLET CONCENTRATION. INLET NO_x CONCENTRATION ~ 1,500 ppm; CATALYST: SUPPORTED NICKEL; GAS MIXTURE: $NO-N_2-He$.

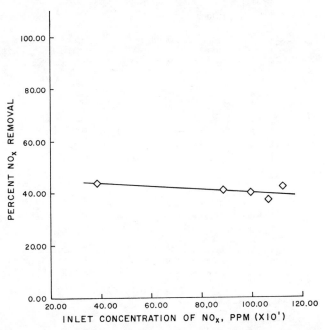

FIGURE 3. EFFECT OF INLET NO_x CONCENTRATION ON CATALYTIC NO_x REMOVAL. CATALYST: SUPPORTED NICKEL; GAS MIXTURE: $NO-N_2-He$; SPACE VELOCITY: 21,000 hr^{-1}.

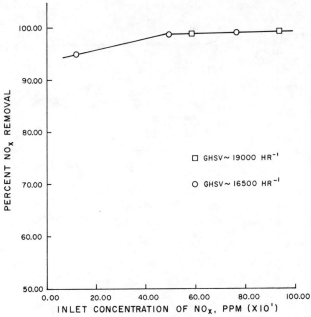

FIGURE 5. CONVERSION AS A FUNCTION OF INLET NO_x CONCENTRATION IN THE PRESENCE OF CARBON MONOXIDE ON SUPPORTED NICKEL CATALYST. GAS MIXTURE: $NO-N_2-CO$.

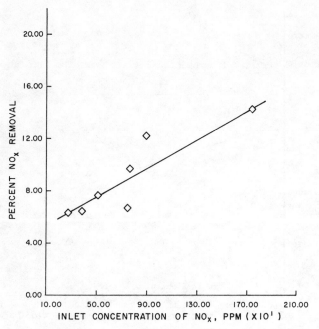

FIGURE 6. NO$_x$ REMOVAL AS A FUNCTION OF INLET CONCEN-
TRATION ON SUPPORTED COBALT CATALYST, AT VARIOUS SPACE
VELOCITIES. GAS MIXTURE: NO-N$_2$-He.

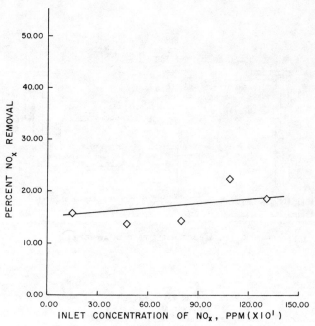

FIGURE 8. UNCATALYZED THERMAL DECOMPOSITION OF NO$_x$.
GAS MIXTURE: NO-N$_2$-He ; AVERAGE GAS VELOCITY: ~2200 m/hr.

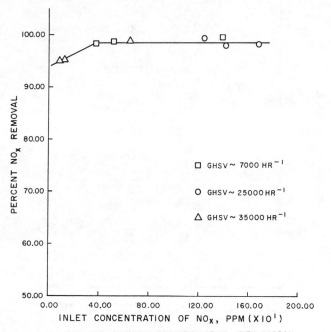

FIGURE 7. CHEMICAL REDUCTION OF NO$_x$ WITH CARBON
MONOXIDE ON SUPPORTED COBALT CATALYST.   GAS MIXTURES :
NO-N$_2$-He-CO AND NO-N$_2$-CO.

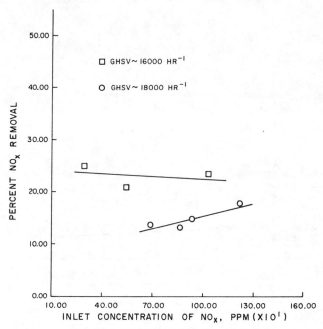

FIGURE 9. CATALYTIC ACTIVITY OF PLAIN ZIRCONIA SUPPORT.
GAS MIXTURE: NO-N$_2$-He.

AN EVALUATION OF SO₂ CONTROL SYSTEMS FOR STEAM GENERATORS AT CALIFORNIA OIL FIELDS

PEDCo Environmental, Inc., has conducted a review of currently available SO₂ control technology as it applies to steam generators at thermally enhanced oil recovery (TEOR) sites in California. This paper summarizes the evaluation study.

In areas with concentrated TEOR activity, present and proposed air emissions regulations were reviewed. A total of 12 available SO₂ control (flue gas desulfurization) processes were assessed, including the sodium-based processes currently being used at California TEOR sites. All the FGD processes were compared at a common design basis. Intermediate-level designs and detailed cost estimates were provided for most of these processes.

The Department of Energy (DOE) contracted this review because TEOR steam production predominantly utilizes fuel containing sulfur levels that produce significant sulfur dioxide emissions. The review performs two functions: 1) it will assist the oil companies as they make decisions concerning SO₂ control system implementation, and 2) it will aid DOE in planning future R&D needs.

AVINASH N. PATKAR

and

SAMIR P. KOTHARI

PEDCo Environmental, Inc.
Cincinnati, Ohio 45246

The purpose of this paper is to present a summary of work performed on a DOE contract that assessed currently available SO_2 control technology and its applicability to thermally enhanced oil recovery (TEOR) operations in California. PEDCo Environmental, Inc., evaluated various flue gas desulfurization (FGD) processes to be considered for use in the California TEOR industry.

The Department of Energy (DOE) has mounted aggressive programs to achieve energy independence because of the economic and strategic problems of reliance on imported oil. Environmental regulations heavily affect the energy industry and DOE programs and hence the attainment of national energy supply goals. The impact of sulfur dioxide (SO_2) emission regulations on the thermally enhanced oil recovery (TEOR) industry in California is an example of such a situation.

Regulatory Background

According to a 1977 California Department of Conservation report, the San Joaquin Valley Air Basin accounted for approximately 80 percent of all TEOR operations in California. Of this portion, 99 percent of the TEOR operations are in Kern County. Therefore, the California Air Resources Board (CARB) has focused regulations on Kern County.

By 1975, approximately 12 million barrels of crude oil were burned annually to generate steam for TEOR operations in Kern County. The burning of this oil, with an average sulfur content of 1.15 percent, resulted in average SO_2 emissions of 130 tons/day. Over the past several years, emissions from TEOR operations have increased significantly. The CARB estimates that by the end of 1978, TEOR operations were generating SO_2 emissions at a rate of about 250 tons/day (based on the assumption that generators operated 80 percent of the time).

These facts and the opinion that the Kern County Air Pollution Control District has not sufficiently dealt with the problem of SO_2 emissions led CARB to impose SO_2 emissions limitations on Kern County steam generators and boilers. The regulations apply to units with heat inputs of at least 15 million Btu/h and were scheduled to take effect in October 1979.

The regulations specify that new sources (i.e., those with construction permits dated on or after February 21, 1979) may emit no more than 0.25 lb sulfur/million Btu input (about 0.24 lb SO_2/million Btu input). For typical crude oil (with a heating value of 150,000 Btu/gal and sulfur content of 1.15 percent), the regulations

S. P. Kothari is now with HDR, Inc., Pensacola, Florida.

0065-8812-81-3804-0211-$2.00
© The American Institute of Chemical Engineers, 1981

require about 90 percent SO_2 removal for a new steam generator. The current regulations of the Kern County require disposal of liquid/solid waste in sites specified as Class I.

Objectives and Contents

The report on the aforementioned work should prove valuable to the oil companies in evaluating trade-offs between various processes, selecting a process, preparing bid specifications, evaluating bids, selecting a vendor, performing an in-house design if desired, and initiating research and development work.

The 12 available FGD processes were examined: sodium carbonate, ammonia, conventional limestone, Chiyoda T-121 limestone, lime, double alkali, dry scrubbing, citrate, sodium sulfite (Wellman Lord), magnesium oxide, carbon adsorption (BF/FW), and copper oxide sorption (Shell/UOP). The first seven are nonregenerable and the last five are regenerable. A regenerable process is defined here as one that both regenerates reagent and produces a salable byproduct.

The paper presents an overview of current FGD activity at California TEOR sites, provides outlines of the FGD processes, and compares the various processes with respect to advantages/disadvantages.

The paper discusses the evaluation approach on a common design basis and compares the FGD system costs according to process area, such as gas handling, feed preparation, SO_2 absorption, and waste handling. It also presents important considerations for process selection, as well as conclusions/recommendations.

FGD SYSTEMS IN CALIFORNIA OILFIELDS

Many FGD systems are used in the California oilfields, and many more are planned for the near future. As of October 1979, there were 79 sodium-based systems operating on 183 steam generators with an overall capacity of 2637 MW_t (9×10^9 Btu/h). By 1982, about 210 systems are expected to operate on nearly 620 generators with an overall capacity of 8204 MW_t (28×10^9 Btu/h.) Table 1 provides detailed information about all units at TEOR sites (1).

The types of companies involved range from a small independent operator using one scrubber on one generator to a company like Getty Oil using 11 scrubbers on a total of 88 generators. The types of scrubbers being used range from a mutually controlled simple eductor design to a fully instrumented tray Getty Oil using 11 scrubbers on a total of 88 generators. The types of scrubbers being used range from a mutually controlled simple eductor design to a fully instrumented tray tower absorber with a complete array of auxiliary equipment. The majority of the companies have purchased scrubber systems from system suppliers while two large companies, Getty and Mobil, have chosen a course of keeping their FGD projects totally in house, from system design to equipment installation and startup.

Most of the currently operating systems are using a similar chemical approach to the problem of SO_2 absorption. Using either sodium hydroxide or sodium carbonate (soda ash) as a makeup reagent, currently operating systems are consistently removing at least 90 percent of the SO_2 in the generator flue gas. On an industry-wide basis, systems' reliability has been high, averaging much higher than that of FGD systems in utility powerplant applications.

This generally successful record is primarily due to the fact that these TEOR scrubbers are utilizing soluble sodium compounds for SO_2 removal. In contrast, over 90 percent of the U.S. utility power plant FGD capacity is using slurried calcium carbonate (limestone) or calcium hydroxide (slaked lime). The average TEOR FGD system controls flue gas equivalent to that of about a 15-MW unit, while the average utility FGD system controls emissions from a boiler generating 400 to 500 MW.

EVALUATED PROCESSES

This section outlines the 12 FGD processes designed for application to TEOR operations.

Sodium Carbonate Process

This is one of the most widely used processes at the TEOR sites. The sodium carbonate process differs from the sodium hydroxide process only in the source of sodium makeup. The major process steps are: quenching and SO_2 absorption, sodium makeup, and waste disposal.

The flue gas is transported by a forced draft fan to a quencher where it is cooled to its dewpoint by recirculating liquor. A small bleed stream is continuously taken from the quencher loop to purge particulates and chlorides.

A fresh sodium carbonate solution (4 percent by weight) is prepared in a mix tank by addition of soda ash from silo and service water. The fresh solution is mixed with the recirculation liquor. The flue gas enters a four-stage tray absorber. The SO_2 is absorbed in a countercurrent mode by recirculation liquor added at the top of the absorber. The gas then passes through a chevron-type mist eliminator, which is washed intermittently by the service water, and finally leaves the system through a stub stack. A constant bleed-off stream is taken to a bleed storage tank and pumped to a waste disposal site.

Ammonia Process (2,3)

The ammonia process employs ammonical liquor containing 25 wt. percent ammonia for SO_2 absorption. The major process steps are similar to those in the sodium carbonate process. The major difference is the type of absorber used, which is a horizontal absorber with three packed beds. The quencher stage with sprays is separated from the rest of the absorber by a chevron-type vertical mist eliminator.

The bleed stream from the first stage is a waste stream that is stored in a large closed tank and may be hauled away to a nearby lined pond.

A common problem in the ammonia process is the formation of blue haze, a nondispersing thick plume of submicron particles (0.2 to 0.5 μm in diameter) of ammonium sulfite/bisulfite/sulfate that scatter light near the wavelength of the color blue. The separation of quencher stage from absorber stages by vertical mist elimination helps reduce the possibility of blue haze formation.

Limestone (Conventional) Process

The limestone process is widely used in the utility industry. The process consists of three major steps: quenching and SO_2 absorption, limestone feed, and sludge dewatering. The makeup reagent is ground limestone fed through a conveyor to a recirculation tank. The quenching and absorption loops are separated by a liquid/gas bowl separator above the quencher. The absorber consists of two spray stages with a small packed section.

A bleed stream of spent slurry from the quencher is pumped to a hydroclone. The concentrated underflow from the hydroclone is fed to a rotary vacuum filter, which produces

a filter cake containing 60 wt. percent solids to be hauled to a sludge pond.

Lime Process

Like the limestone process, the lime process depends on a calcium-based reagent and is widely used in the utility sector. The major differences are slaking requirements for lime and type of absorber used. The major steps are similar to the limestone process.

A horizontal absorber with four stages of sprays is employed for SO_2 absorption. The dewatering step is identical to that in the limestone process.

Double Alkali Process

Like sodium-based processes, the double alkali process removes SO_2 by scrubbing the flue gas with a sodium-based solution. In the double alkali process, however, sodium is regenerated from the bleed stream, and chemical costs are minimized. The process uses lime and soda ash as raw materials both in a fine powdered form.

The process comprises three major steps: quenching and SO_2 absorption, feed preparation, and regeneration and waste handling. The SO_2 absorber is a tower with three flexi-trays. For regeneration and waste handling, the effluent from the lime reactor, containing 2 wt. percent solids, is concentrated to 30 wt. percent solids slurry in a thickener. Makeup soda ash is added to the thickener at a rate proportional to losses in the filter cake. The thickener underflow is fed to the vacuum filter, which produces filter cake containing 60 percent solids to be hauled to a lined pond. The thickener overflow containing regenerated sodium sulfite is pumped to the top of the absorber.

Dry Scrubbing Process

The dry scrubbing process differs from other nonregenerable processes because it generates dry waste product rather than a wet sludge. The dry scrubbing process evaluated in the study employs trona (a monoclinic mineral containing sodium bicarbonate and sodium carbonate) as an alkali absorbent. The main steps in the process are SO_2 absorption with particulate removal and feed preparation.

The SO_2-laden flue gas contacts an atomized scrubbing liquor in a spray dryer and vaporizes all the water; the sodium salts flow out with the gas stream. The solids in the gas are collected in a fabric filter downstream of the spray dryer. The solids are pneumatically conveyed to a storage silo.

Citrate Process

This is a regenerable FGD process in which SO_2 is removed from flue gas by the buffering action of a citrate solution. The U.S. Bureau of Mines has developed a process that regenerates the citrate solution with hydrogen sulfide and produces elemental sulfur as a byproduct. The evaluated process, however, regenerates the citrate solution by steam stripping and produces liquid SO_2 as the salable byproduct.

The main steps in the citrate process are quenching and SO_2 absorption, regeneration, purge treatment, and byproduct handling. In a packed absorber, the flue gas and citrate solution are contacted in a countercurrent manner. The amount of citrate solution required for SO_2 absorption depends mainly on the temperature of the absorption operation and SO_2 content of the flue gas.

Sodium Sulfite (Wellman Lord) Process

The sodium sulfite (Wellman Lord) process removes SO_2 from flue gas with a sodium sulfite scrubbing solution. Dry, concentrated SO_2 from the regeneration area can be converted into a marketable product, such as liquid SO_2, sulfuric acid, or elemental sulfur. The evaluation assumes that liquid SO_2 is produced as a byproduct because of the small quantity involved and low costs.

The process uses very little raw material because regeneration reduces soda ash requirements. The regeneration section, however, requires substantial energy in the form of steam for conversion of sodium bisulfite to sodium sulfite. The process consists of three major steps: quenching and SO_2 absorption, purge treatment, and regeneration. The absorber has five trays, each connected to a separate recirculation loop. The absorption reaction is reversed in the evaporator by heating the sodium bisulfite; as a result, sodium sulfite is regenerated.

Magnesium Oxide Process

In this regenerable process, magnesium-based slurry is contacted with flue gas containing SO_2. The SO_2 is absorbed and forms an insoluble product, which is removed from the scrubbing solution by a centrifuge. The product

is dried and shipped to a central facility for the recovery of its sulfur value as sulfuric acid and for regeneration of the reagent, magnesium oxide (MgO). The regenerated MgO is recycled back to the FGD system.

The magnesium oxide process has been commercially demonstrated, and its chemistry is well developed. The process poses no significant solid waste or wastewater problems and can be adapted to multiple generators in the oilfields, so that only one centrally located regeneration facility is required.

Carbon Adsorption (BF/FW) Process

The carbon adsorption process is a dry regenerable process using activated charcoal for removal of SO_2 from flue gas by adsorption. The process was developed by Bergbau-Forschung, CmbH., in West Germany and its licensee in the U.S. is Foster Wheeler. The process requires large amount of equipment and has high energy consumption. It produces a salable byproduct sulfur.

Copper Oxide Sorption (Shell/UOP) Process

This dry regenerable process removes SO_2 from flue gas by copper oxide sorption and produces a salable byproduct sulfur. The process is expensive, and energy requirements are great. Although parts of the process system have been tested, a fully integrated system has yet to be demonstrated.

Table 2 compares various advantages and disadvantages of the evaluated FGD processes.

EVALUATION APPROACH AND COST ESTIMATES

The steps involved in the evaluation of each process were as follows:

Review of available literature: an information base was established from in-house files and published literature; additional information required to complete the task was obtained by contacting knowledgeable sources; FGD process flow diagrams were prepared for oilfield applications.

Examination of development and application: whether the process has been demonstrated on a steam generator with an output of 50 million Btu/h or more was examined along with any special requirements

Calculation of material balances: the flow rate and composition of each process stream were defined for the common design basis; the basic determining factor was the amount of reagent required to remove the necessary amount of SO_2; individual and overall material balances were calculated on the basis of key design requirements, such as operating liquid-to-gas ratio, recirculating pH, pressure drop across the absorber, and bleed-off requirements

Sizing of equipment items and determination of energy requirements: individual material balances and design conditions were used to size process equipment items and determine operating energy requirements

Estimation of capital investment and annual cost requirements: for meaningful economic comparisons, a cost basis was established

Identification of design considerations: critical design aspects of the process were identified and important operating variables were noted

Assessment of environmental considerations: waste disposal requirements were examined and problems pertinent to TEOR site application were assessed

Table 3 presents the common design basis for the evaluated FGD systems.

Cost Estimates

This subsection compares capital investment and annual cost estimates for 11 of the 12 FGD processes outlined in Section 3. The estimates for the copper oxide process are not available, but are of the same order-of-magnitude as estimates for the carbon adsorption process.

The accuracy range of cost estimates for all processes except carbon adsorption is +30 percent to -10 percent. For carbon adsorption, the accuracy range is +40 percent to -25 percent.

Preliminary analysis of the carbon adsorption and copper oxide processes indicated that they are unsuitable for TEOR application because of technical complexity and excessive energy and capital investment requirements.

For these processes, only order-of-magnitude cost estimates were made. The other FGD processes were evaluated in more detail.

Table 4 summarizes capital investment and annual cost estimates for nonregenerable processes by process area. Unit capital investments range from \$20.7 to \$50.9/Nm^3/h of flue gas treated (\$13.13 to \$32.32/scfm). Unit net annual costs range from \$7.11 to \$11.04/m^3 (\$1.12 to \$1.74/ bbl) of oil burned. Estimated capital investment is lower for the Chiyoda T-121 limestone process than for the conventional limestone process because the former requires only an interim pond with a life of 2 years whereas the latter requires a pond with a life of 20 years. The estimated cost of the SO_2 absorption section is significantly higher for the dry scrubbing process than for other processes because of the fabric filter and spray dryer with atomizer. Estimated annual costs are lower for the ammonia process than for the sodium carbonate process because the former requires less expense for reagent to remove a unit weight of SO_2 than the latter.

Table 5 summarizes capital investment and annual cost estimates for regenerable processes by process area. Unit capital investments range from \$42.1 to \$66.2/Nm^3/h (\$26.67 to \$41.96/scfm) of flue gas treated. Unit net annual costs range from \$8.3 to \$15.7/m^3 (\$1.31 to \$2.47/bbl) of oil burned. The capital investment and annual cost requirements of a centrally located sulfuric acid plant used in the magnesium oxide process are assumed to be distributed equally among eight identical FGD systems. The contributions to such a plant are included in the capital investment and annual cost estimates for the process. Net annual costs are lower for the sodium sulfite (Wellman Lord) process than for the citrate process because soda ash makeup costs less than citric acid makeup.

PROCESS SELECTION

Many complex, interrelated factors affect selection of an FGD process for a specific oilfield site. This section provides three typical scenarios and a selection procedure that consists of preliminary screening and process rating. The procedure eliminates some types of FGD processes from further consideration. Even after completion of the procedure, however, additional information may be required to select a particular process for a specific application.

Scenarios

Three scenarios are considered in this study (4). Scenario 1 assumes the following conditions:

New generators 14.7 MW$_t$ each
Firing of low-sulfur oil (1.1 percent sulfur)
Availability of lime and limestone
Availability of large pond area for sludge disposal
Strict water effluent restrictions
No market for sulfur compounds
Possibility of manifolding

Scenario 2 assumes the following conditions:

New generators, 14.7 MW$_t$ each (50 x 10^6 Btu/h each)
Firing of low-sulfur oil (1.1 per cent sulfur)
Availability of soda ash, ammonia, and limestone
No transportation facilities for sludge disposal
Restriction of wastewater ponding to clay-lined ponds
No market for liquid SO$_2$, sulfur, or sulfuric acid
Possibility of market for ammonium compounds and gypsum
No possibility of manifolding

Scenario 3 assumes the following conditions:

New generators, 14.7 MW$_t$ each (50 x 10^6 Btu/h each)
Firing of high-sulfur oil or medium-sulfur coal
Availability of trona and soda ash in large quantities
Restricted availability of land for waste disposal
Possibility of marketing liquid SO$_2$
Possibility of manifolding

Preliminary Screening

Preliminary screening of FGD processes involves evaluation of four main factors: raw material requirements, end product disposal/sales, performance requirements, and plantsite considerations.

Raw Material Requirements. The raw material required by the 12 FGD systems include ash or caustic, ammonia, lime, limestone, magnesium oxide, citric acid, activated carbon, hydrogen, and copper. The availability of these reagents should be reviewed critically.

End Product Disposal/Sales. The waste products produced by the 12 FGD systems include sodium sulfite/bisulfite/sulfate liquor and calcium sulfite/sulfate sludge. Salable products include liquid SO$_2$, sulfuric acid, elemental sulfur, gypsum, and ammonium sulfite/bisulfate/sulfate liquor. Disposal practices required should be reviewed with respect to land availability and current and future regulations; in addition, marketability of the byproducts should be ascertained.

Performance Requirements. All processes should be evaluated on the basis of several requirements of performance. An SO$_2$ removal efficiency of 90 to 95 percent is required for new installations in the oilfields. In addition, the potential for fly ash and especially nitrogen oxides control is desirable and should be examined. Finally, each process should have been demonstrated on a commercial-scale operation with a heat output of at least 300 x 10^6 Btu/h and with boilers firing high- and low-sulfur oil.

Plantsite Considerations. A retrofit application (i.e., installation of an FGD system on an existing generator) usually does not pose space constraints. Manifolding of several generators, however, may be possible and should be considered. If the generators are isolated, manifolding is impossible. This situation effectively rules out all regenerable processes, because they are prohibitively expensive on a "one-on-one" basis.

Applications. The preliminary screening procedure can be applied to each scenario. The preliminary screening would favor certain FGD processes and rule out other processes for a particular scenario.

Process Rating (5)

After preliminary screening, process rating indicates the most feasible processes. Process rating does not identify a single best process, but ranks the applicable processes.

Criteria for Process Rating. Table 6 presents criteria for process rating in oilfield applications. The categories covered include system status, raw material, byproducts/waste, energy needs, pollution control, and costs.

Most point ratings (R's) are self-explanatory. Under system status, for example, if an FGD process has been applied to units with low average capacity, it is assigned a high point rating. Except for costs, R's vary from 0 to 3. For costs, they range from 3 to 10.

A weighing factor (W) is included for the owner's use. This factor enables an oilfield engineer to determine the factors most important to a particular site. For example, marketability of sludge produced is not important for Scenario 1 and is weighted low. Conversely, it is very important for Scenarios 2 and 3 and is weighted high. Except for costs, W's vary from 0 to 10. Because costs are important irrespective of scenario, they are weighted with an 8 or 10.

Applications. The criteria for process rating can be applied to the FGD processes favored for each scenario by preliminary screening. The point ratings are assigned to these processes for each criterion and are multiplied by the weighing factors for the scenario to determine scores for overall importance.

CONCLUSIONS/RECOMMENDATIONS

Conclusions

Several conclusions can be drawn from the elevation study based on individual plantsite requirements. The ammonia process is about 10 percent more cost-effective than the sodium-based processes now used in the oilfields if the absorber bleed stream can be disposed of in a lined pond or sold for fertilizer use. The lime, conventional limestone, double alkali, and Chiyoda T-121 limestone processes are very competitive with sodium-based processes, if sludge disposal in a lined pond is permissible. Although economically unattractive for oil-fired generators, dry scrubbing is preferable to all other processes if the generators fire low-sulfur coal because it offers high particulate removal efficiency as well as SO_2 control. The citrate and sodium sulfite (Wellman Lord) processes are feasible only if several generators can be manifolded, if liquid SO_2 can be marketed, and if waste disposal regulations are very stringent. The magnesium oxide process is feasible only if a centrally located facility for reagent regeneration and sulfuric acid production is

possible and if waste disposal regulations are very stringent.

Recommendations

Because of the potential growth of the enhanced oil recovery (EOR) industry and increase in FGD systems in the California oilfields, modifications of the present FGD system should be considered. In addition, short-and long-term research projects are needed to demonstrate pollution control systems that simultaneously remove ash, SO_2, and NO_x. Various projects are recommended in the study.

Although the sodium carbonate process has demonstrated high reliability and SO_2 removal efficiency, absorbent costs and possible restrictions on current waste disposal practices may create problems. Thus, sodium sulfate production, regeneration with lime, and electrodialytic regeneration should be considered.

Except for the blue haze problem, the ammonia process is economical and offers the potential for producing a fertilizer byproduct. Ammonium sulfate production needs to be evaluated for technical feasibility and product marketability. The selective catalytic reduction technology (SCR) developed for NO_x control uses ammonia (NH_3) as the reductant. The SCR technology should be evaluated for oilfield application in combination with the ammonia FGD process.

The magnesium oxide process produces a solid or liquid magnesium sulfite/sulfate product, which can be easily transported to a processing plant. The practicability of a central sulfuric acid plant in the oilfields area that would receive the product from several facilities using the MgO process should be evaluated.

Because the California Air Resource Board is likely to require NO_x emission control significantly greater than combustion modification techniques, a pilot-plant study of dry SO_2 removal by carbon adsorption integrated with dry NO_x removal by the SCR method is recommended.

As oil prices increase, it may become cost-effective for the oil companies to burn low-sulfur Wyoming coal in the steam generators. Fluidized bed combustion (FBC), which involves burning fuel in an air-fluidized bed of limestone or dolomite, may then become feasible. The FBC can meet stringent SO_2

emission standards. An FBC unit with particulate control device should be compared with the use of a conventional oil-fired steam generator and an FGD system.

LITERATURE CITED

1. Tuttle, J., et al., "EPA Industrial Boiler FGD Survey, First Quarter 1979," EPA-600/7-79-067b (April 1979).

2. Johnstone, M. F., et al., "Recovery of SO_2 from Waste Gases, Equilibrium Vapor Pressure Over Sulfite-Bisulfite Solutions," Ind. & Eng. Chemistry, 1(30): 101-109 (1938).

3. Johnstone, M. F., "Recovery of SO_2 from Waste Gases," Ind. & Eng. Chemistry, 12(29):1396-98 (Dec. 1937).

4. Patkar, A. N., et al., "An Evaluation of FGD Processes for Application to California Oilfields," final report, DOE/ET/12088-1, Section 7 (April 1980).

5. Torstrick, R. L., et al., "Flue Gas Desulfurization Applicability Study, NATO-CCMS Study, Phase 1,2," preliminary report (May 1979).

TABLE 1 (continued)

Oil company	Location	Sulfur in fuel oil, %	SO$_2$ removal efficiency, %	FGD unit supplier	No. of FGD units	FGD status	FGD startup date	No. of generators under control	Total capacity under control, MW$_t$ (10^6 Btu/h)
Petro Lewis	Bakersfield	1.08	95	Thermotics	1	a	Feb. 1979	1	14.7 (50)
Rainbow	Bakersfield	1.1	90	Thermotics	1	a	Feb. 1979	1	5.3 (18)
Santa Fe Energy	Bakersfield	1.5	96	FMC	1	a	Sept. 1979	7	90 (310)
	Various locations			Not selected	10-20	c	By mid-1981	79	570 (3000)
Shell	Oildale	1.1		Not selected	2 or 3	c	Early 1980	12	176 (600)
	Taft	1.1		Not selected	1	c	Early 1980	2	29 (100)
Sun Production	Oildale	1.2	94	C-E Natco	1	a	Sept. 1979	1	14.7 (50)
	Fellows	1.4	94	C-E Natco	1	a	Sept. 1979	1	14.7 (50)
Texaco	San Ardo	1.7	95	Ducon	3	a	By Sept. 1979	9	131 (450)
	San Ardo	1.7	73	Ceilcote	29	a	By mid-1974	29	435 (1560)
	San Ardo	1.7	95	Not selected		e		12	174 (600)
	Santa Maria	3.45		Not selected	1	c	Jan. 1980	4	58 (200)
	Fellows	1.5		Not selected		c	Mar. 1980		
Union	Guadalupe	3.0	96	Heater Technology	2	a	1978	2	29 (100)
	Guadalupe	3.0	97	Koch Engineering	1	a	1978	1	7.3 (25)
	Guadalupe	3.0	96	Thermotics	1	a	1978	1	7.3 (25)
	Guadalupe	3.0	90+	Koch Engineering	4	c	By June 1980	4	29 (100)
	Guadalupe	3.0	90+	Heater Technology	12	c	By June 1980	12	133 (475)
	McKittrick	1.39	90+	Anderson 2000	1	c	Apr. 1980	1	14.7 (50)
	North Belridge[f]	1.39	90+	Anderson 2000	1	c	May 1980	1	14.7 (50)
	Coalinga	1.39	90+	Anderson 2000	2	c	Mar. 1980	2	29 (100)
	Various locations			Not selected	30	c	By 1981	100	1450 (5000)

a Operational.

b Under construction.

c Planning or considering SO$_2$ control.

d Shut down.

e Five units are operational, 10 units will be restarted in 1980, and 13 are permanently shut down.

f Leased.

TABLE 2. ADVANTAGES AND DISADVANTAGES OF FGD PROCESSES

FGD Process	Advantages	Disadvantages
Sodium carbonate	No scaling or plugging, small liquid-to-gas ratio (L/G), high SO$_2$ removal, wide availability from vendors	Possible major waste disposal problem, expensive reagent
Ammonia	No scaling or plugging, small L/G, reagent cheaper than soda ash, possible market for by-product, adaptability for simultaneous removal of SO$_2$ and nitrogen oxides (NO$_x$)	Control of blue fume needed
Limestone (conventional)	Much operating experience, minimal scaling with dual-loop design, least expensive alkali	Large quantities of solid waste, large L/G, corrosion/erosion at wet/dry interfaces
Limestone (Chiyoda T-121)	Gypsum produced for stacking or marketing, inexpensive alkali, minimal scaling and plugging	Large pressure drop, available only in demonstration stage in United States
Lime	Small pressure drop through horizontal scrubber, less scaling than with limestone process, cheap alkali	Large quantities of solid waste, large L/G, corrosion/erosion at wet/dry interface
Double alkali	All advantages of sodium carbonate process with greatly reduced reagent costs and liquid waste problem	Large quantities of solid waste, high costs
Dry scrubbing	Easily disposable dry waste product, very small L/G, high particulate removal	Economic feasibility only when particulate removal is required as well as SO$_2$ removal
Citrate	Minimal waste product, small L/G, marketable liquid SO$_2$ produced	Gas cooling below saturation point, reheat required, high costs
Sodium sulfite	Same as for Citrate process, many successful installations	Same as for Citrate process
Magnesium oxide	Less scaling potential than with lime process, marketable sulfuric acid produced, minimal waste product	High costs, regeneration facility required with acid plant
Carbon adsorption	Dry process, marketable sulfur produced, potential for NO$_x$ control	Very high costs, energy intensive process

TABLE 3.   DESIGN BASIS FOR A PROPOSED FGD SYSTEM
IN A CALIFORNIA OILFIELD APPLICATION

Steam generator characteristics:

Duty	87.9 MW_t (300 million Btu/h) output, total 6 generator bank, each at 14.7 MW_t (50 million Btu/h)
Steam rate	128,122 Kg/h (282,380 lb/h), 80 percent quality
Excess air	15 percent
Load factor	90 percent

Fuel oil properties:

Sulfur, wt. percent	1.14
H.H.V.	42,240 J/m^3 (150,000 Btu/gal)

Flue gas characteristics at FGD system inlet:

Total flow rate	62.96 m^3/h at 260°C (133,800 acfm at 500°F)
Molecular weight	29.05 (wet basis), 30.44 (dry basis)
Carbon dioxide, vol percent	12.2
Water vapor, vol percent	11.2
Oxygen, vol percent	2.6
Sulfur dioxide	600 ppmv

Emission regulation:

25.77 ng/J (0.06 lb sulfur/million Btu) input, maximum, which is approximately equal to 51.55 SO_2 ng/J (0.12 lb/million Btu) input

TABLE 4. COMPARISON OF CAPITAL AND ANNUAL COSTS OF NONREGENERABLE PROCESSES
(thousands of dollars)

	Sodium carbonate	Ammonia	Limestone	Limestone (Chiyoda)	Lime	Double alkali	Dry scrubbing
CAPITAL INVESTMENT[a]							
Direct investment							
Feed preparation	37.91	17.63	29.13	40.06	42.49	58.26	15.70
Gas handling	42.90	64.00	41.97	71.50	57.54	42.90	62.30
SO_2 absorption	114.64	124.47	219.09	211.82	211.83	126.97	800.40
Waste handling	28.27	31.28	46.19	42.62	42.57	88.72	14.20
Reheat	-	-	-	-	-	-	-
Purchased equipment cost	223.72	242.38	336.38	366.00	354.43	316.85	892.60
Pond construction	94.20	74.03	151.05	33.00	144.10	140.60	-
Other direct costs	261.75	283.59	393.57	428.22	414.69	459.44	533.77
Total direct investment	579.67	600.00	881.00	827.22	913.22	916.89	1426.37
Indirect investment							
Engineering fees	86.95	90.00	132.15	124.08	136.98	137.53	213.96
Other indirect costs	173.90	180.00	264.30	248.17	273.96	275.07	427.91
Total indirect investment	260.85	270.00	396.45	372.25	410.94	412.60	641.87
Contingency	126.08	130.50	191.62	179.93	198.62	199.42	310.24
Total capital investment	966.60	1000.50	1469.07	1379.40	1522.78	1528.91	2378.48
Unit investment, $/scfm	13.13	13.59	19.96	18.74	20.69	20.77	32.32
Unit investment, $/Nm^3/h	20.7	21.4	31.5	29.6	32.63	32.5	50.9
ANNUAL COSTS[a]							
Direct costs							
Raw materials	202.20	138.60	73.49	70.72	74.83	85.58	75.10
Water	3.14	3.33	2.79	2.36	2.79	2.79	2.80
Electricity	44.10	44.10	65.40	96.90	64.86	48.10	88.40
Waste hauling	23.80	18.40	7.24	7.24	7.24	7.24	4.00
Operating labor and supervision	60.44	60.44	60.44	60.44	60.44	75.56	60.50
Other utilities	-	-	-	-	-	-	19.50[b]
Subtotal	333.68	264.87	209.36	237.66	210.17	219.27	250.30
Maintenance and repairs	50.43	52.20	76.65	71.97	79.46	79.77	124.10
Total direct costs	384.11	317.07	286.01	309.63	289.63	299.04	374.40
Indirect costs							
Fixed costs	57.16	59.16	86.86	81.57	90.04	90.40	140.63
Overheads	60.06	60.77	70.55	68.67	71.69	81.78	89.53
General expenses	90.37	93.55	137.36	128.98	142.38	142.95	222.38
Total indirect costs	207.59	213.48	294.77	279.22	304.11	315.13	452.54
Credits							
Byproduct sales revenue	-	-	-	(17.50)	-	-	-
Heat recovery	-	-	-	-	-	-	-
Total credits	-	-	-	(17.50)	-	-	-
Net annual cost	591.70	530.55	580.78	571.35	593.74	614.17	826.94
Unit annual cost							
$/bbl oil burned	1.24	1.12	1.22	1.20	1.25	1.29	1.74
$/m^3 oil burned	7.88	7.12	7.75	7.63	7.95	8.19	11.06

[a] California oilfield location, mid-1979 costs. Design basis: 300×10^6 Btu/h (total heat output), six steam generators, 1.14% sulfur in oil, 90% SO_2 removal, 90% load factor, pond life of 20 years, interim pond life of 2 years for limestone (Chiyoda T-121) process. A dash indicates that item does not apply.

[b] Bag replacement cost.

TABLE 5. COMPARISON OF CAPITAL AND ANNUAL COSTS OF REGENERABLE PROCESSES
(thousands of dollars)

	Citrate	Wellman Lord	Magnesium oxide	Carbon adsorption[a]
CAPITAL INVESTMENT[b]				
Direct investment				
Feed preparation	28.05	10.15	26.45	57.5
Gas handling	50.88	54.54	41.00	122.6
SO_2 absorption	267.08	243.14	174.13	310.6
Effluent handling	-	-	171.91	-
Reheat	4.32	3.17	-	-
Purge treatment	7.89	2.90	-	-
Regeneration	109.16	176.14	c	265.3
Purchased equipment cost	467.38	490.04	413.49	756.0
Pond construction	32.00	21.70	-	-
Other direct costs	677.70	710.56	599.56	1096.2
Total direct investment	1177.08	1222.30	1013.05	1852.2
Indirect investment				
Engineering fees	176.56	183.34	151.96	277.8
Other indirect costs	353.12	366.69	303.92	555.7
Total indirect investment	529.68	550.03	455.88	833.5
Contingency	256.01	265.85	220.34	402.9
Total capital investment	1962.77	2038.18	2251.27[c]	3088.6
Unit investment, $/scfm	26.67	27.69	30.59	41.96
Unit investment, $/Nm^3/h	42.1	43.7	48.2	66.2
ANNUAL COSTS[b]				
Direct costs				
Raw materials	169.41	20.81	47.20	328.6
Water	1.37	1.00	2.79	2.5
Electricity	85.50	109.80	61.32	168.0
Operating labor and supervision	90.67	90.67	75.56	90.6
Other utilities	204.78	166.28	35.76	173.7
Subtotal	551.73	388.56	222.63	763.4
Maintenance and repairs	102.40	106.34	88.14	161.1
Total direct costs	654.13	494.90	310.77	924.5
Indirect costs				
Fixed costs	116.06	120.52	99.89	182.6
Overheads	100.80	102.37	85.13	124.3
General expenses	183.52	190.57	157.94	288.8
Total indirect costs	400.38	413.46	342.96	595.7
Credits				
Byproduct sales revenue	(112.40)	(102.18)	(52.06)	(46.3)
Heat recovery	(195.60)	(182.28)	-	(300.0)
Total credits	(308.00)	(284.46)	(52.06)	(346.3)
Net annual cost	746.51	623.90	744.07[c]	1173.9
Unit annual cost				
$/bbl oil burned	1.57	1.31	1.57	2.47
$/m^3 oil burned	9.9	8.3	9.9	15.7

[a] Order-of-magnitude estimate of costs.

[b] California oil field location, mid-1979 costs. Design basis: 300 x 10^6 Btu/h (total heat output), six steam generators, 1.14% S in oil, 90% SO_2 removal, 90% load factor, purge disposal pond life of 20 years. A dash indicates that item does not apply.

[c] Includes additional capital and annual costs of $562,000 and $142,400 contributing to central regeneration facility respectively.

TABLE 6. CRITERIA FOR PROCESS RATING

	Rating, R			Weighting factor, W		
	Low	Medium	High	Scenario 1	Scenario 2	Scenario 3
System status						
Number of system installed	1	2	3	5	5	5
Average capacity of systems installed	3	2	1	5	5	5
Operating lime in United States	1	2	3	5	5	5
Process controllability	1	2	3	8	8	8
Mechancial complexity	3	2	1	9	9	9
Average availability	3	2	1	10	10	10
Maintenance requirements	3	2	1	10	10	10
Space requirement	3	2	1	5	5	5
Raw materials						
Availability of sorption agent	1	2	3	10	10	10
Freshwater requirement	3	2	1	4	4	4
Reducing agent requirement	3	2	0	4	4	9
Fuel oil requirement	3	1	0	4	4	8
Requirements for other raw materials	3	2	1	8	8	8
Byproducts/Waste						
Marketable quality	0	2	3	0	9	9
Marketability	0	2	3	0	9	9
Sludge produced	3	2	1	4	8	10
Wastewater produced	3	2	1	10	9	10
Land requirement	3	2	1	4	7	10
Energy needs						
Steam (reheat and process)	3	2	1	8	8	10
Electricity	3	2	1	6	6	6
Pollution control						
SO_2 removal	0	2	3	10	10	10
Ash removal	0	2	3	6	6	10
NO_x removal	0	2	3	8	8	8
Halides removal	1	2	3	6	6	6
Secondary air emissions	2	0	0	6	6	6
Costs						
Capital investment	10	7	3	8	8	8
Annual costs	10	7	3	10	10	10

FIBER BEDS FOR CONTROL OF SULFURIC ACID MIST EMISSIONS

An experimental investigation of dioctyl phthlate (DOP) mist removal from an air stream with glass fiber beds has been performed. Fiber size and bed velocities were varied so mist elimination by both inertial impaction and Brownian diffusion could be studied. Mist loading and particle size distributions were determined by cascade impactor sampling.

By careful analysis of the DOP-air results, the performance of fiber beds for elimination of H_2SO_4 mist could be predicted. Fiber beds operating in the inertial impaction regime are suitable for control of emissions from 98% H_2SO_4, contact processes. For oleum processes, however, fiber beds operating in the Brownian diffusion regime are required for proper emission control.

T. L. HOLMES

Koch Engineering Company, Inc.
Wichita, Kansas 67208

SCOPE

For the last 20 years, fiber beds have been successfully used for removal of submicron mists from process gas streams. The three primary mechanisms for mist removal are: direct interception of the mist by the fibers; inertial impaction of the mist onto the fibers; and collision of smaller mist particles with the fibers due to Brownian diffusion. Since significant velocity profile disturbances are present in a fiber bed, due to the interference of adjacent fibers, a generalized theory for quantitatively characterizing aerosol deposition is presently not available. In view of this, design of fiber bed mist eliminators is typically based upon experimental results obtained with pilot or full scale systems.

Several fundamental studies of the mist or particle collection which occurs when a gas stream flows over a single fiber have been performed. Ranz and Wong (1) have considered dust and smoke collection by interception and inertial impaction. Langmuir in Rodebush et al. (2) has considered the collection of oleic acid mist by Brownian diffusion. Results of these studies are of significant value in the planning of fiber bed experiments and in the analysis of the results.

Although both total mist loading and opacity of gas streams leaving a fiber bed may be important for proper emission control, only mist loading control is considered here. The objective of this study is to present a method of analysis of pilot-scale, mist collection efficiency data which will permit the general design of fiber beds for mist elimination to be readily accomplished. By employing a fractional collection efficiency technique, mist collection is characterized as a function of particle size. If for some reason the inlet particle size distribution to the fiber bed is altered, the resulting performance of the mist eliminator can be predicted. This fractional collection efficiency technique along with the results of the fundamental single fiber studies previously mentioned, permit experimental data obtained with one aerosol-gas system to be used to design a mist eliminator for an entirely different aerosol-gas system. In view of this, the number of experiments required for the design of mist eliminators can be significantly reduced.

CONCLUSIONS AND SIGNIFICANCE

Pilot scale experiments have been performed with vertical cylinder glass fiber beds for removing a 1.4 micron mass median size dioctyl phthalate (DOP) mist from a room temperature air stream. These beds were constructed of two concentric 304 stainless steel screens which contained the glass fiber. By careful selection of gas velocity and fiber size, mist collection by

0065-8812-81-4161-0211-$2.00

predominately inertial impaction (the "Impaction Cylinder") and predominately Brownian diffusion (the "BD Cylinder") could be studied. Mist loading and particle size distributions at the inlet and outlet of the radial flow fiber beds were determined with a University of Washington "Mark 3" cascade impactor. Fractional collection efficiency data, in turn, were calculated from the particle size distribution curves.

By correcting the pilot-scale, air-DOP results for fluid physical properties, pressure drop and fractional collection efficiencies for both IC and BD fiber beds were predicted for removing 98% H_2SO_4 mist from gas streams in the contact sulfuric acid process. Results predicted for the impaction cylinder were found to be in excellent agreement with measured performance of a 1010 kg/sec, 98% H_2SO_4 plant.

The predicted fractional collection efficiency data for IC and BD cylinders were used to calculate the performance of these fiber beds for removal of H_2SO_4 mist from the tail gas stream from the final absorbing tower in bright sulfur, dark sulfur, bound sulfur, and 20% oleum contact plants. The impaction cylinder was found to yield satisfactory mist collection performance for bright and dark sulfur burning and bound sulfur plants. For the 20% oleum plant, however, the BD cylinder was required to control the mist emission to below 7.5×10^{-5} kg mist per kg of 100% sulfuric acid produced.

BACKGROUND

The two main types of processes in use for H_2SO_4 manufacture are the contact process and the older lead chamber process. As discussed in the EPA Guideline Series (3), in 1971, contact process plants accounted for over 97% of the U.S. production capacity. A discussion of the contact process is given below.

In a typical contact process, molten sulfur is burned in air to yield a SO_2 gas stream. The SO_2 is then catalytically oxidized to SO_3 in a 4-stage converter. The SO_3, in turn, is then absorbed into a 98% H_2SO_4 stream in a packed absorbing tower. A number of variations of this process are possible depending upon the source of SO_2 gas. If molten sulfur containing less than 0.1% hydrocarbon is burned, the contact process is termed a "bright sulfur" process (see Brink, et al. (4)). If the molten sulfur contains

over 0.1 to 0.2% hydrocarbon content, the contact process is termed a "dark sulfur" process. When the source of SO_2 gas comes from an ore roaster or from burning spent acid, the resulting process may be termed a "bound sulfur" process (see EPA Guideline Series (3)). The acid produced in a contact process is typically 98% H_2SO_4. If, however, additional SO_3 is added to the 98% circulating acid in a packed oleum tower, the resulting process is termed an "oleum" process. Additional discussion of each of these processes is given by Shreve and Brink (5).

The most common sulfuric acid process contains a single SO_3 absorption tower. Since the oxidation of SO_2 to SO_3 is a reversible reaction, increased conversion of SO_2 can be obtained by removing some of the SO_3 from the gas stream between the 3rd and 4th stages of the 4 stage converter. The removal is accomplished by passing the gas stream through a packed absorption tower (interpass absorber). The scrubbed gas is then reheated and admitted to the 4th stage of the converter. Additional discussion of dual absorption plants is given by Duros and Kennedy (6).

The particle size distribution of the acid mist formed in a H_2SO_4 plant is a strong function of the type of contact process used. As discussed by Brink, et al. (4), 0.01 to 10 micron sulfuric acid mists are formed in one of two ways, which are: 1) gas phase reaction between water vapor and SO_3 to form H_2SO_4 vapor which condenses at normal absorber temperatures, and 2) shock cooling of converter gases as they are admitted to the oleum and 98% H_2SO_4 absorption towers which permits H_2SO_4 vapor to condense. A drying tower is used in the contact process to remove water vapor from the air which is subsequently admitted to the sulfur burner and converter. This tower is a packed absorber where the air is contacted with either 93 or 98% H_2SO_4 (see EPA Guideline Series (3)). With dark sulfur processes, hydrocarbons burn in the sulfur burner to form CO_2 and water vapor which provides another source of moisture to form H_2SO_4 mists. In a single absorption plant, normal final absorbing tower temperatures are in the 350 to $360°K$ range. If oleum is being produced, however, converter gases are cooled to 311 to $333°K$. The severe quench in this oleum tower provides another source of acid mist which will pass through the final absorption tower and the auxiliary

mist elimination equipment if it is not prop-
erly designed.

Typical particle size distribution
curves of acid mist formed in the various
processes are shown in Figure 1. Data from
three sources was used in the construction of
these curves (York and Poppele ($\underline{7}$), Duros
and Kennedy ($\underline{6}$), and the EPA Guideline Series
($\underline{3}$)). For single absorption plants, curve 1
is typical of the mist at the mist eliminator
inlet of the final absorbing tower in a 20 to
25% oleum process; curve 2 is typical of such
inlet mist in a bound sulfur or dark sulfur
burning process and curve 3 is a conservative
estimate of such inlet mist in a bright sulfur burning process. For dual absorption
plants, curves 1 and 2 are typical of the
mist at the primary absorption tower mist
eliminator inlet and curve 3 is typical of
the mist at the mist eliminator inlet of the
final absorbing tower.

As discussed in the EPA Guideline Series
($\underline{3}$), both electrostatic precipitators and
fiber bed contactors are used for elimination
of acid mist from H_2SO_4 process gas streams.
Although the gas-phase pressure drop of an
electrostatic precipitator is low, typically
0.0254 m of water, relatively high capital
and operating cost of this device have hin-
dered its use for H_2SO_4. Most H_2SO_4 plants
use fiber beds for mist elimination. The 3
types of fiber bed mist eliminators currently
in use are: the vertical cylinder, the vertical panel, and horizontal dual pads. The
vertical cylinder is constructed of 2 concentric 316 SS screens which contain the fiber
packing. The typical dimensions of this cylinder are 0.610 m OD by 3.05 m long. These
cylinders are installed vertically on a tube
sheet, typically in the top of an absorption
tower. The vertical panel contains two parallel 316 SS or Alloy 20 screens which contain the fiber packing. Typical dimensions
of this panel are 0.47 m by 1.35 m. These
panels are installed vertically in a polygon
arrangement in the top of an absorption tower.
With both the cylinder and the panel, gas and
mist flow cocurrently through the packing.
The collected liquid drains from the downstream side of the fiber bed, through a seal
leg, and back into the absorption tower. The
horizontal dual pad mist eliminator, however,
contains two horizontal fiber beds mounted
one above the other in the top of an absorbing
tower. Liquid mist collected with these pads
falls countercurrently to the rising gas
stream back into the absorption tower. The
fiber packing used in these mist eliminators

is either glass or fluorocarbon.

The mechanisms of mist collection by
fibrous beds are interception, inertial impaction, and Brownian diffusion. Typical
gas velocities employed with vertical panel
and horizontal dual pad mist eliminators are
in the 2.03 to 3.05 m/sec range and the primary mist collection mechanisms for these
beds are interception and impaction. With
vertical cylinder beds, gas velocities are
usually maintained in the 0.1 to 0.2 m/sec
range, as discussed in the EPA Guideline
Series ($\underline{3}$), and the predominant mist collection mechanism is Brownian diffusion. Based
upon single fiber studies, the mist collection efficiency is known to increase with a
dimensionless separation number, ψ. For
interception and impaction, these separation
numbers, as formulated by Ranz and Wong ($\underline{1}$),
are:

$$\text{Interception:} \quad \psi = \frac{D_p}{D_c} \qquad (1)$$

$$\text{Impaction:} \quad \psi = \frac{C \rho_p\ U_0\ D_p{}^2}{18\ \mu\ D_c} \qquad (2)$$

Langmuir in Rodebush, et al. ($\underline{2}$), has considered mist collection by Brownian diffusion. A separation number for this mist
collection mechanism is:

$$\psi = D_0\ t \qquad (3)$$

If the characteristic time in Equation (3)
is assumed equal to the time required for
the mist particle to travel one fiber diameter, the separation number for Brownian
diffusion becomes:

$$\psi = \frac{D_0}{U_0\ D_c} \qquad (4)$$

Equations (1) through (4) are most useful in
selecting proper fiber size and gas velocity
for mist collection. Mist collection efficiency is enhanced by selecting small fiber
diameters. This efficiency is enhanced by
high gas velocity with beds operating in the
inertial impaction regime, while it is reduced by increasing gas velocity, hence particle velocity, for beds operating in the
Brownian diffusion regime. Due to the complex problem of characterizing fluid flow
through fiber beds, formulation of a quanti-
tative theory for predicting mist collection
efficiency is a difficult task. In view of
this, mist collection efficiency is usually
measured experimentally. Such an experimental program for characterizing mist

collection efficiency and the application of the results to the design of H_2SO_4 mist eliminators are discussed below.

EXPERIMENTAL EQUIPMENT AND PROCEDURE

A simplified schematic of the fiber bed pilot plant is shown in Figure 2. This system contains: a DOP aerosol generator; a room air prefilter which also serves as a mist-air mixing chamber; a 0.61 m diameter by 1.57 m high test chamber which houses a small scale version of the vertical cylinder fiber bed previously discussed; and a gas metering and flow section which contains an orifice flow meter, an automatic butterfly valve and an ID fan.

In operation of this system, DOP mist generated by means of 2-fluid nozzles in the aerosol generator, is mixed with filtered room air and admitted to the test chamber. The mist and air flow radially through the glass fiber bed in the vertical test cylinder. Both inertial impaction and Brownian diffusion were studied with this test cylinder by selecting suitable fiber sizes and bed velocities. The liquid DOP, which collects in the fiber bed, drains by gravity from the inner screen to the seal pot located below the test chamber. The cleaned gas, which typically contains very little DOP mist, is drawn out the top of the test chamber and through the orifice meter by means of an ID fan or blower.

A desired inlet mist loading was set by choosing the proper number of 2-fluid nozzles in the aerosol generator. The appropriate gas velocity was set with the automatic butterfly valve. Depending upon the inlet mist loading and gas velocity selected, 50 to 150 hours of continuous operation were required to achieve steady state conditions. Since the liquid content of the fiber bed increases with time, ie, liquid accumulates until steady state is reached, the fiber bed pressure drop increases with time. To maintain a constant gas flow rate through the fiber bed, the automatic butterfly valve opens slightly as the wet bed pressure drop increases. Steady state conditions were assumed to be reached when the wet bed pressure drop was constant with time.

Once steady state was reached, the wet fiber bed pressure drop was measured with a U-tube manometer. Inlet and outlet mist loading and particle size distributions were measured with a University of Washington, 7 stage cascade impactor. Although the inlet and outlet gas streams were sampled in succession rather than simultaneously, repeated sampling of the inlet gas stream revealed that fairly constant mist loading and particles size distributions were maintained at the inlet to the fiber bed. In all cases isokinetic sampling was performed following the recommended sampling procedures discussed by Harris (8).

EXPERIMENTAL RESULTS

The cascade impactor used in this study contains 7 stages and a final filter. Each stage contains multiple nozzles which subdivide the gas into a number of equal width or diameter jets which impact on collection plates located immediately below the stages. Mist particles are collected on these plates by inertial impaction. In view of this, the starting point for analysis of cascade impactor data is Equation (2) with U_0 replaced with the jet velocity V_j and the collector diameter replaced by the jet diameter, D_j. The resulting equation is:

$$\psi = \frac{C \rho_p V_j D_p^2}{18 \mu D_j} \qquad (5)$$

The jet velocity is calculated from the total sampler volumetric flow rate Q as follows:

$$V_j = \frac{4Q}{\pi D_j^2 N} \qquad (6)$$

Now defining a mist cut diameter, D_{50}, as the diameter of mist particles which are collected with 50% efficiency on a given stage and assuming a 50% separation number of 0.145 for circular jets as discussed by Pilat et al. (9), the following equation results for calculating particle size:

$$D_{50} = \left(\frac{2.05 \mu D_j^3 N^{\frac{1}{2}}}{C \rho_p Q}\right) \qquad (7)$$

Now since

$$C = 1 + \frac{2\lambda}{D_{50}} \left(1.257 + 0.4\, e^{-1.1 \frac{D_{50}}{2\lambda}}\right) \qquad (8)$$

Equations (7) and (8) represent 2 equations which may be solved for the 2 unknowns D_{50} and C. Since these 2 equations cannot be solved analytically, computer program MIST-SIZE was written to solve for C and D_{50} with a half-internal root location technique.

Particle size distribution data calculated from inlet and outlet cascade impactor sampling results obtained with the impaction cylinder, fiber bed pilot plant for the DOP-room temperature air system are shown in Figure 3. The curves shown in this figure, are an eyeball fit of the data. Also shown on this figure are predicted results for the 98% H_2SO_4, 355°K absorber tail gas system. The sulfuric acid curves were generated by using the density of 98% H_2SO_4 in Equation (7). As discussed by York and Poppele ($\underline{7}$), the effects of gas phase physical properties on collection efficiency are negligible, when predicting absorber tail gas results from room temperature air data. As can be seen from Equation (7), increases in viscosity and volumetric flow rate tend to virtually cancel each other over the 294 to 355°K temperature range of interest.

From the particle size distribution curves in Figure 3 and experimentally measured mist loading values, the fractional collection efficiency of the impaction cylinder for removing 98% H_2SO_4 mist could be calculated. Such efficiencies were calculated in a ± 0.05 micron particle size range around each particle size of interest. These predicted results are shown as data points in Figure 4. Actual performance data for impaction cylinders removing 98% H_2SO_4 mist from the final absorber tail gas in a 1010 kg/sec plant are shown as the curve in this figure. Excellent agreement between actual and predicted mist collection performance is observed. Measured gas-phase pressure drop for the plant impaction cylinder mist elimination system is 0.274 m of water. The measured pressure drop for the pilot scale impaction cylinder is 0.224 m of water. Correcting this value for fluid physical properties and differences in mist loading the predicted pressure drop for this absorber tail gas mist eliminator is 0.279 m of water.

In a manner similar to that discussed above for the impaction cylinder, fractional collection efficiency data for a vertical cylinder fiber bed operating in the Brownian diffusion regime were determined. Results for the room temperature air-DOP system and the 355°K absorber tail gas-98% H_2SO_4 system are shown in Figure 5. The data points were calculated from pilot-scale experimental data and the curves shown are an eyeball fit of these data. The fractional collection efficiency is seen to start decreasing as the particle size decreases below one micron characteristic of the inertial impaction

mechanism (see Figure 4). A minimum in the fractional efficiency curve is reached at 0.5 to 0.2 microns and then it increases at even smaller particle sizes due to Brownian diffusion. The glass fiber size used in the BD cylinder and the bed velocities considered are significantly less than those used with the impaction cylinder. Measured BD pressure drop for the room temperature air-DOP system is 0.056 m of water. Predicted pressure drop for this BD cylinder operating on 180°F absorber tail gas is 0.061 m of water.

ANALYSIS AND APPLICATION OF RESULTS

The fractional efficiency curves shown in Figures 4 and 5 are felt to be typical of the performance of fiber bed mist eliminators currently in use in contact sulfuric acid processes. Indeed, by using glass fibers with diameters smaller than those considered in the BD cylinder test work, an ultra high efficiency fiber bed contactor can be designed which exhibits a minimum in the fractional efficiency vs. particle size curve which is above 99.5%. Since these ultra high efficiency contactors are required only in special circumstances, eg., plants producing 40 to 60% oleum, they are not considered in the following discussion.

To more fully characterize the performance of glass fiber beds operating in inertial impaction and Brownian diffusion regimes for removing sulfuric acid mist, it is of interest to calculate the overall mist collection performance for various absorber tail gas streams. For simplicity only single absorption contact plants will be considered in these calculations. These calculations are performed for mist elimination from final absorber tail gas streams for the following 3 types of plants: 1) bright sulfur burning; 2) bound sulfur and dark sulfur burning; and 3) 20% oleum producing plants. As discussed in the EPA Guideline Series ($\underline{3}$), typical uncontrolled mist emissions for 98% H_2SO_4 plants are 2×10^{-3} kg mist/kg of 100% H_2SO_4 produced and such emissions for oleum plants are typically 5×10^{-3} kg mist/kg of 100% H_2SO_4 produced. With this inlet loading basis, the inlet particle size distribution curves in Figure 1, and the fractional efficiency curves the overall mist elimination performance of these fiber beds can be calculated.

In the calculation of overall collection efficiency, the inlet mist loading in

each of a number of small mist size ranges
was determined and using the fractional collection efficiency at the midpoint of each
diameter size range the fractional penetration was determined. The overall penetration
of mist through the fiber bed is simply the
sum of the fractional penetration values.
The overall collection efficiency, in turn,
is calculated as 1 minus (penetration/inlet
mist loading). These calculations are summarized in Table 1 for the impaction cylinder
and in Table 2 for the Brownian diffusion
cylinder. A summary of overall performance
results for both types of fiber beds is given
in Table 3.

As discussed in the EPA Guideline Series
(3), the recommended maximum emission of
H_2SO_4 mist from new plants is 7.5×10^{-5} kg
mist/kg of 100% H_2SO_4 produced. This exit
mist loading is calculated as kg 100% H_2SO_4
mist/kg of 100% H_2SO_4 produced. From the results in Table 3, fiber beds operating in the
inertial impaction regime are suitable for
control of normal emissions from bright sulfur, dark sulfur, and bound sulfur plants.
For the plant producing 20% oleum, however,
the fiber bed operating in the Brownian diffusion regime is required for proper control
of mist emissions. From a capital investment
standpoint, the impaction cylinder is preferred for 98% H_2SO_4 mist elimination, since
30 times less fiber bed area is required to
process a given volumetric flow rate of gas.

ACKNOWLEDGEMENT

This work was performed at the Research
and Development Laboratory in Wichita, Kansas.
The permission of Koch Engineering Company,
Inc., to publish this paper is greatly appre-
ciated.

NOTATION

BD	= Brownian diffusion fiber bed
C	= Cunningham slip correction factor
D_c	= diameter of collector in Equation (1)
D_{50}	= cut size for a cascade impactor stage
D_o	= diffusion coefficient in Equation (3)
D_p	= diameter of a mist particle
IC	= inertial impaction fiber bed
N	= number of jets per cascade impactor stage
P	= mist penetration through fiber bed
Q	= gas volumetric flow rate through cascade impactor
t	= characteristic time in Equation (3)
U_o	= particle velocity
V_j	= gas velocity through cascade impactor jets

Greek Letters

Δ	= net mist loading in a given particle size range
η	= fractional collection efficiency
λ	= mean free path of gas molecules
μ	= gas viscosity
ψ	= separation number
ρ	= density of a mist particle

Subscripts

c	= collector
50	= cut size
j	= jet
p	= particle
o	= particle

LITERATURE CITED

1. Ranz, W. E., and J. B. Wong, Ind. & Eng. Chem., 44 (6), 1371 (1952).

2. Rodebush, W. H., I. Langmuir, and V. K. LaMer, "Filtration of Aerosols and the Development of Filter Materials," OSRD No. 865, Serial No. 353 (1942).

3. EPA Guideline Series, "Control of Sulfuric Acid Mist Emissions from Existing Sulfuric Acid Production Units," EPA450/277019, QAQPS No. 1.2078 (1977).

4. Brink, J. A., Jr., W. F. Burggrabe, and L. E. Greenwell, Chem. Eng. Prog., 64 (1), 82 (1968).

5. Shreve, R. N., and J. A. Brink, Jr., Chemical Process Industries, Chapter 19, 4th ed., McGraw-Hill, New York, N.Y. (1977).

6. Duros, D. R., and E. D. Kennedy, Chem. Eng. Prog., 74 (9), 70 (1978).

7. York, O. H., and E. W. Poppele, Chem. Eng. Prog., 66 (11), 67 (1970).

8. Harris, B. D., "Procedures for Cascade Impactor Calibration and Operation in Process Streams," Environmental Protection Technology Series, EPA-600/2-77-004 (1977).

9. Pilat, M. J., D. S. Ensor, and J. C. Bosch, Atmospheric Environment, 4, 671 (1970).

Table 1. Overall Mist Collection Efficiency of the Impaction Cylinder

Type of Contact Process

D_p Microns	$1-\eta$ %	Bright Sulfur Burning			Dark or Bound Sulfur			20% Oleum		
		Cum %<	Δ, %	$P \times 10^4$	Cum %<	Δ, %	$P \times 10^4$	Cum %<	Δ, %	$P \times 10^4$
3.0	0.00	62.00			76.00			92.50		
2.5	0.22	44.50	17.50	3.850	72.80	4.00	0.880	90.30	2.20	0.484
2.0	0.22	17.00	27.50	6.050	63.20	9.60	2.112	85.90	4.40	0.968
1.5	0.08	6.20	10.80	0.864	44.10	19.10	1.528	76.30	9.60	0.768
1.2	0.30	3.30	2.90	0.870	24.20	19.90	5.970	65.20	11.10	3.330
1.0	1.30	2.30	1.00	1.300	14.10	10.10	13.130	53.10	12.10	15.730
0.9	2.50	1.90	0.40	1.000	10.40	3.70	9.250	46.00	7.10	17.750
0.8	3.70	1.56	0.34	1.258	7.50	2.90	10.730	36.90	9.10	33.670
0.7	5.60	1.29	0.27	1.512	5.20	2.30	12.880	27.30	9.60	53.760
0.6	8.50	1.04	0.25	2.125	3.50	1.70	14.450	17.50	9.80	83.300
0.5	18.50	0.84	0.20	3.700	2.20	1.30	24.050	10.40	7.10	131.350
0.4	33.10	0.70	0.14	4.634	1.20	1.00	33.100	5.30	5.10	168.810
0.3	48.10	0.52	0.18	8.658	0.59	0.61	29.341	2.10	3.20	153.920
0.2	70.60	0.37	0.15	10.590	0.22	0.37	26.122	0.61	1.49	105.194
0.0	85.40	0.00	0.37	31.598	0.00	0.22	18.788	0.00	0.61	52.094
				78.009			202.331			821.128

Overall Collection Efficiency, %   99.22   97.98   91.79

Table 2. Overall Mist Collection Efficiency of the BD Cylinder

Type of Contact Process

D_p Microns	$1-\eta$ %	Bright Sulfur Burning			Dark or Bound Sulfur			20% Oleum		
		Cum %<	Δ, %	$P \times 10^4$	Cum %<	Δ, %	$P \times 10^4$	Cum %<	Δ, %	$P \times 10^4$
3.0	0.00	62.00			76.80			92.50		
2.5	0.005	44.50	17.50	0.088	72.80	4.00	0.020	90.30	2.20	0.001
2.0	0.01	17.00	27.50	0.275	63.20	9.60	0.096	85.90	4.40	0.044
1.5	0.03	6.20	10.80	0.324	44.10	19.10	0.573	76.30	9.60	0.228
1.2	0.06	3.30	2.90	0.174	24.20	19.90	1.194	65.20	11.10	0.666
1.0	0.20	2.30	1.00	0.200	14.10	10.10	2.020	53.10	12.10	2.420
0.9	0.38	1.90	0.40	0.152	10.40	3.70	1.406	46.00	7.10	2.698
0.8	0.71	1.56	0.34	0.241	7.50	2.90	2.059	36.90	9.10	6.461
0.7	1.13	1.29	0.27	0.305	5.20	2.30	2.599	27.30	9.60	10.848
0.6	1.57	1.04	0.25	0.393	3.50	1.70	2.669	17.50	9.80	15.386
0.5	2.00	0.84	0.20	0.400	2.20	1.30	2.600	10.40	7.10	14.200
0.4	2.37	0.70	0.14	0.332	1.20	1.00	2.370	5.30	5.10	12.087
0.3	2.58	0.52	0.18	0.464	0.59	0.61	1.574	2.10	3.20	8.256
0.2	2.69	0.37	0.15	0.404	0.22	0.37	0.995	0.61	1.49	4.008
0.0	2.63	0.00	0.37	0.973	0.00	0.22	0.579	0.00	0.61	1.604
				4.725			20.754			78.917

Overall Collection Efficiency, %   99.95   99.79   99.21

Table 3. Summary of Fiber Bed Performance on Various H_2SO_4 Mists

Type of Contact Process

Item	Bright Sulfur Burning	Dark or Bound Sulfur	20% Oleum
Typical Final Absorber Mist Eliminator Inlet Mist Loading, kg/kg 100% H_2SO_4	2.00×10^{-3}	2.00×10^{-3}	5.00×10^{-3}
Mass Median Size of Inlet Mist, microns	2.63	1.60	9.60×10^{-1}
Max. Allowable Mist Emission, kg/kg 100% H_2SO_4	7.50×10^{-5}	7.50×10^{-5}	7.50×10^{-5}
I. Performance of Impaction Cylinder			
Overall mist collection efficiency, %	99.22	97.98	91.79
H_2SO_4 Mist Emission, kg/kg 100% H_2SO_4	1.50×10^{-5}	4.00×10^{-5}	4.10×10^{-4}
Pressure Drop, m of H_2O	2.80×10^{-1}	2.80×10^{-1}	3.00×10^{-1}
Relative Velocity	1.00	1.00	1.00
Relative Bed Area	1.00	1.00	1.00
II. Performance of BD Cylinder			
Overall mist collection efficiency, %	99.95	99.79	99.21
H_2SO_4 Mist Emission, kg/kg 100% H_2SO_4	1.00×10^{-6}	4.00×10^{-6}	3.95×10^{-6}
Pressure Drop, m of H_2O	6.10×10^{-2}	6.10×10^{-2}	6.55×10^{-2}
Relative Velocity	3.3×10^{-2}	3.3×10^{-2}	3.3×10^{-2}
Relative Bed Area	30.00	30.00	30.00

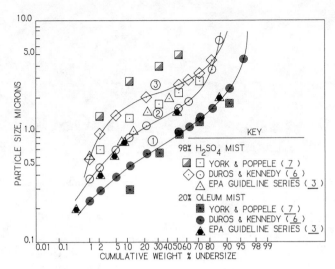

Figure 1. Typical mist eliminator inlet H₂SO₄ mist size distributions.

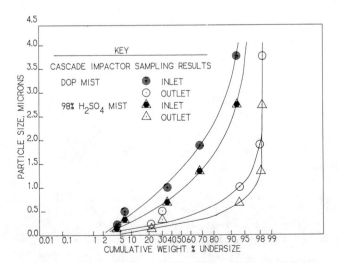

Figure 2. Simplified schematic of fiber bed pilot plant.

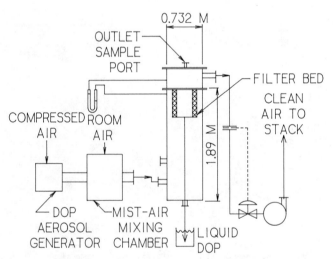

Figure 3. Particle size distributions of DOP and H₂SO₄ mist.

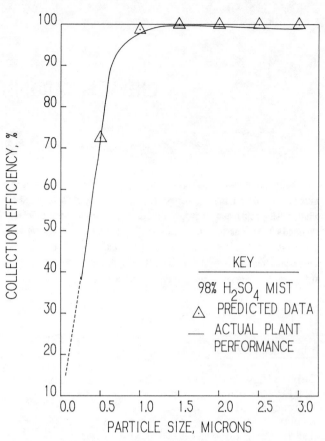

Figure 4. Fractional mist collection efficiency of the impaction cylinder.

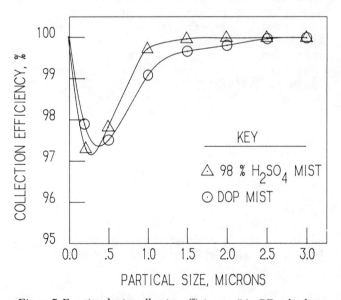

Figure 5. Fractional mist collection efficiency of the BD cylinder.

GENERATION OF HIGH CONCENTRATION WAX AEROSOLS

A. S. DAMLE

and

R. MAHALINGAM

Department of Chemical Engineering
Washington State University
Pullman, Washington 99164

High concentration solid wax aerosols in the inhalable particle size range (<15 μm) were generated by atomizing molten wax with a stream of compressed air, followed by dilution using cold clean air supply. Aerosol outputs in the range 0.025-0.05 g/min were obtained with air pressures of 15 to 55 psig (2×10^5 to 4.8×10^5 pascals) and atomizer air flow rates of 1.5 to 4.5 l/min. The geometric mean particle diameter decreased with increasing air pressure up to 35 psig (3.45×10^5 pascals), after which it remained steady at 1.8 μm. The aerosol output, however, increased steadily with increasing air pressure.

Introduction

Aerosol systems required for experimental applications vary widely in necessary characteristics. A variety of methods are used to produce such aerosols under laboratory conditions. A detailed review of the commonly used laboratory techniques was recently made available by Raabe (1976). Laboratory aerosol research is primarily concerned with inhalation studies and with the basic understanding of aerosol science. For such studies, generation of low concentration, near monodisperse fine aerosols is needed. For industrial research, however, high concentration, polydisperse solid aerosols are preferred in order to evaluate full-scale equipment performance and aerosol system interactions. Heavy dust loadings have the advantage of minimizing sampling times involved. Redispersion of solid dust is usually employed to obtain a desired high concentration aerosol stream. This technique produces much coarser aerosols and requires adequate control of humidity conditions to obtain reproducibility (Grassel, 1976). This might be satisfactory where the emphasis is on efficient

total mass collection; however, the present-day emphasis on fine particulate control and the possibility of establishment of a new source standard for inhalable particles, less than 15 μm (Drehmel, 1979), create a need for a high concentration, solid, fine aerosols generation system.

Most of the laboratory techniques to produce fine solid aerosols involve generation of droplets of a solution or suspension, followed by solvent vaporization. The evaporation stage decreases the net output of solid aerosols considerably. A condensation method was recently used (Prodi, 1972) to generate solid wax aerosols; however, due to low vapor pressures involved, the output was low. In the present work a simple method is described to generate high concentration solid aerosols by atomization of molten wax followed by solidification of the wax droplets (Damle, 1979).

Experimental Setup

The laboratory setup basically consisted of an air-supply system, liquid atomizer, diluter and a Kr-85 neutralizer. A schematic diagram of the generation system is given in Figure 1. Thermo-Systems, Inc., Model 3073 liquid atomizer was used in the present studies; however, any other liquid atomizer capable of generating fine droplets would be ade-

---

A.S. Damle is presently with the Research Triangle Institute, Research Triangle Park, North Carolina 27709.

0065-8812-81-3798-0211-$2.00

quate. The TSI model is similar to the Collison nebulizer as described by May (1973). Plain white technical-grade paraffin wax (M.P. ~ 45°C) was used in the present work. A heating tape was wound around the stainless-steel liquid container of the atomizer to keep the wax in molten state and its temperature was controlled to 75°C by a regulated power supply.

Compressed air supplied by a GMAT oil-less compressor, A, was passed through an absolute filter, B, a pressure regulator, C, and a silica gel dryer, D. Part of this compressed air went to the heated liquid atomizer, E. The regulated air pressure determines the air flow through the atomizer nozzle and the flow rate was monitored by a calibrated rotameter. The freshly generated aerosol stream was mixed with the remaining compressed air stream in the dilution chamber, F. The dilution air was regulated by a needle valve and a calibrated rotameter. The wax droplets solidified in the dilution chamber. The aerosol stream was then passed through a Kr-85 deionizer, G, to neutralize any charges generated by the atomizing process. An assembly, H, of a baffle impactor and two cyclones in series with an effective cutpoint diameter of 1 μm was later used to investigate the feasibility of obtaining a finer aerosols stream (d_{max} < 2 μm).

Analyses and Results

With the arrangement used, the particle size distribution primarily depends upon the atomizing air pressure. The aerosol output was analyzed for its size distribution and concentration, with a University of Washington Mark III cascade impactor. The generation system was run at constant atomizing pressures varying between 15 and 55 psig (2×10^5 and 4.8×10^5 Pa) at 5 psi (3.45×10^4 Pa) intervals. Since the atomizing air flow rate varies with the air pressure, the dilution air flow rate was adjusted to obtain a total flow rate of 1.06 m³/hr. To eliminate sampling errors, the total aerosol stream was sent through the impactor. The cutpoint particle diameters for different stages were obtained from calibration curves and thus represent aerodynamic diameter of particles.

The data obtained at various pressures are tabulated in Table 1 and the cumulative aerosol size distributions at 3 different pressures are shown in Figure 2. The percentage values were obtained on mass basis. The average particle size becomes smaller with increasing pressure up to about 35 psig after which the size distribution remains unchanged. Contrary to our expectation, the size distribution is not lognormal but is rather skewed towards smaller size. The d_{50} particle diameter ranges from 2.7 μm at 15 psig to 1.8 μm at 35 psig or higher. As seen from Figure 2, practically all generated particles are in the inhalable size range, being below 15 μm.

Along with the size distribution data, Table 1 includes the atomizer air flow rate, aerosol output, and aerosol concentration before dilution at all atomizing air pressures used. Aerosol output increases with increasing pressure due to the higher atomizer air flow rates; however, the aerosol concentration simultaneously goes down with an apparent leveling off trend at pressures above 35 psig. This is due to the removal of a greater fraction of generated droplets within the atomizer because of the higher velocity of air-droplet jet impacting on the inner wall. As seen from the table the aerosol concentration before dilution ranged from 0.01 to 0.015 g/l with an output of 0.025 to 0.05 g/min.

The stability of the generator is good both for continuous and intermittent operation. The maximum amount of time this generator may be operated continuously depends primarily upon the volume of liquid contained in the atomizer and the aerosol output rate and can be increased by designing a larger liquid container. For a constant wax temperature and atomizing air pressure, the aerosol characteristics are reproducible. Starting from room temperature the desired temperature of molten wax is reached within 15 to 20 minutes. After this the air flow rate is started and the steady-state conditions of aerosols generation are reached within a few seconds.

For applications requiring a larger aerosol output than that obtained by a single nozzle generator, a multiple nozzle system may be designed or alternatively several individual units may be operated in parallel.

The aerosol size distribution may further be modified by using suitable intermediate systems. In this work a baffle impactor and two cyclones in series (Assembly H) were used to remove larger particles. The resulting size distribution for an atomizer air pressure of 35 psig is also shown in Figure 2. As can be seen from this figure, 70 percent of particles were below 1 μm. The fine particle size was verified by observing an aerosol sample under scanning electron microscope. The aerosol stream was sampled through a nucleopore filter paper (pore size 0.08 μm) for this purpose. Typical sample pictures are shown in Figure 3. The maximum time this system could be run continuously was, however, limited to an hour, after which the plugging of aerosol lines tends to alter the size distribution and concentration of aerosols.

Conclusion

The system for generation of solid-wax aerosols as described in this report yields high concentration, polydisperse aerosols in the inhalable size range. The method shows good consistency, reproducibility and stability for intermittent operations. The generator, therefore, is suitable for studies requiring a high concentration, polydisperse aerosols system, for example in laboratory-scale testing of a control concept or a control equipment.

Acknowledgment

Initial support received from Electrical Power Research Institute and subsequent support by Washington State University, Department of Chemical Engineering, are hereby acknowledged.

Literature Cited

1. Raabe, O.G., "The Generation of Aerosols of Fine Particles," in Fine Particles Symposium Proceedings; Edited by B.Y.H. Liu, Minneapolis; Academic Press, New York, 1975.

2. Grassel, E.E., "Aerosol Generation for Industrial Research and Product Testing," in Fine Particles Symposium Proceedings; Edited by B.Y.H. Liu, Minneapolis; Academic Press, New York, 1975.

3. Drehmel, D., "Considerations for Establishing a Standard for Inhaled Particles," Presented at The Second Symposium on Advances in Particle Sampling and Measurement, Daytona Beach, Florida, October 1979.

4. Prodi, V., "A Condensation Aerosol Generator for Solid Monodisperse Particles," Chapter 9, Assessment of Airborne Particles; Edited by Mercer T.T., Morrow, P.E., and Stober, W., Rochester, New York; C.C. Thomas, Publisher, Springfield, Illinois, 1972.

5. Damle, A.S., "Gas-Submicron Particles Separation in a Flowing Liquid Foam Bubble Matrix," Ph.D. Dissertation, Department of Chemical Engineering, Washington State University, Pullman, WA, 1979.

6. May, K.R., "The Collison Nebulizer: Description, Performance and Application," J. of Aerosol Science, Vol. 4, p.235-243, 1973.

7. TSI, Inc., Instruction Manual for Model 3073, 1977.

Table 1. Molten Wax Atomizer Data at Various Pressures

		Pressure (psig)								
		15	20	25	30	35	40	45	50	55
Stage No.	Cutpoint	CPL	CPL	CPL	CPL	CPL	CPL	CPL	CPL	CPL
1	14	100	100	100	100	100	100	100	100	100
2	5.8	87.2	89.2	91.3	93.9	96.1	96.6	96.2	96.4	96.0
3	2.8	64.3	66.8	69.4	71.7	73.4	73.9	74.3	74.5	73.9
4	1.7	35.7	37.6	40.4	44.0	46.4	46.3	45.9	45.8	46.2
5	0.8	16.2	18.0	20.4	22.5	24.4	24.2	24.0	24.1	24.8
6	0.4	10.2	12.0	14.0	16.4	18.2	18.1	17.6	17.8	18.1
Filter	-	0	0	0	0	0	0	0	0	0
Atomizer flow rate l/min		1.69	2.03	2.32	2.71	2.92	3.26	3.67	4.08	4.49
Aerosol output g/min		0.0247	0.0267	0.0288	0.0325	0.0345	0.0372	0.0403	0.0433	0.0488
Aerosol concentration g/l		0.0146	0.0132	0.0124	0.0120	0.0118	0.0114	0.0110	0.0106	0.0108

Sampling flow rate - 17.6 l/min, sampling time - 1 min.

CPL = cumulative percent lower.

Stage cutpoint diameters in microns.

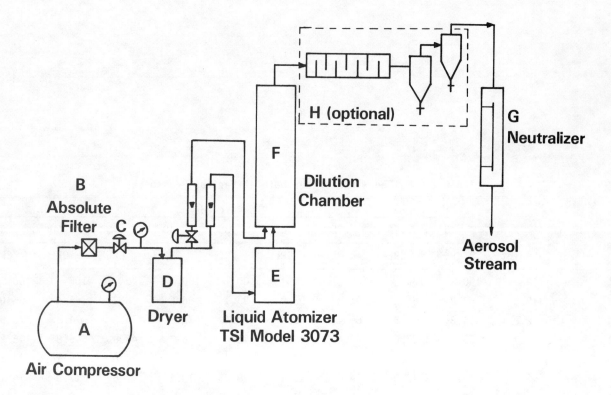

Figure 1. Wax aerosol generation system.

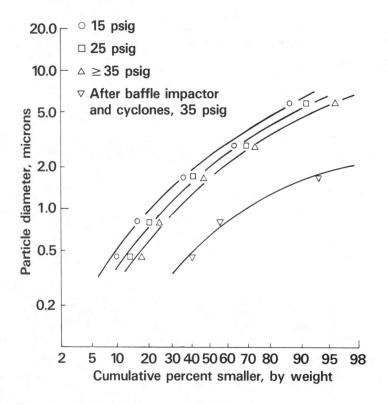

Figure 2. Size distribution for wax aerosol, various atomizing air pressures.

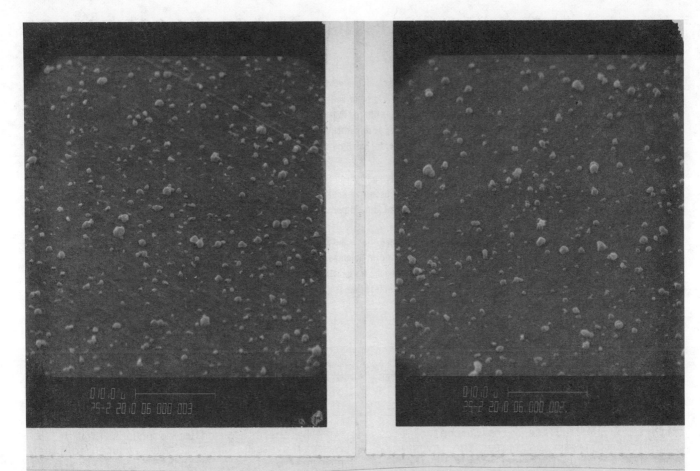

Figure 3. Scanning electron microscope photograph of the wax aerosol samples. (Aerosol stream was passed through a baffle impactor and two cyclones.)

SO₂ REMOVAL USING DRY SODIUM COMPOUNDS

A process for control of SO₂ emission using dry sodium compounds injected into the flue gas duct system of a fabric filter is being developed. The finely ground sodium carbonates are reacted to sulfates or sulfites in the duct and as the gas is filtered through the dust cake. Thus, the fabric filter becomes a device for controlling both particulate emission and SO₂ emission. Minimal additional equipment is necessary for handling and injecting the dry powder.

Tests are being made on a 1.42-m³/s (3000-cfm) electric utility flue gas slipstream using three potential sodium compounds: (1) nahcolite, a naturally occurring sodium bicarbonate, (2) air-dried crude ore (crude trona) from Owens Lake, and (3) a 70-percent pure sodium bicarbonate made from dry lake ore.

This paper reports the results of a series of EPA-sponsored tests to evaluate the SO₂ removal performance of each compound as a function of the stoichiometric ratio, temperature, and filter air-to-cloth ratio. The estimated cost of SO₂ removal using the dry injection of sodium compounds is lower than that of spray drying, which in turn is less than that of wet limestone scrubbing, for a low sulfur coal application.

ERIC A. SAMUEL
DALE A. FURLONG

Envirotech Corp.
Lebanon, Pennsylvania 17042

THEODORE G. BRNA

United States Environmental Protection Agency
Research Triangle Park, North Carolina 27711

and

RONALD L. OSTOP

City of Colorado Springs
Colorado Springs, Colorado 80903

Sodium compounds are reasonably reactive with sulfur dioxide (SO₂) at temperatures typical of coal-fired boiler flue gases. This fact, together with the increasing use of fabric filters for particulate emission control of coal-fired boilers, instigated the investigation of dry sodium compounds for removing SO₂ from flue gas. The possibilities of such a process were suggested by the aluminum industry's success with a dry additive fabric filter collector system for the control of gaseous and particulate fluorides in the aluminum potline effluent.

There have been a number of investigations of ways to remove SO₂ with solid sorbents.[1] The sorbents have been various limestones, dolomites, quicklime, hydrated lime, manganese dioxide, sodium bicarbonate, sodium carbonate, and potassium permanganate. Investigations have confirmed that only sodium carbonate and sodium bicarbonate have good capability for reducing SO₂. To assess candidates for further testing, an investigation was made of the availability of low cost sources of sodium arbonate and bicarbonate prior to the present work. Previous testing had shown the apparent superiority of sodium bicarbonate in the form of nahcolite for reaction with oxides of sulfur and nitrogen (SOₓ/NOₓ) in the dry form.

Other candidate sorbents were evaluated as alternates considering the current availability of nahcolite.

Before using sodium compounds in dry SO₂ removal systems, they must be reduced to fine powders for injection and to increase the surface area for improved chemical reactivity. Generally, powders poassing through 200-mesh screen have permitted excellent SO₂ removal.

Considerable economic incentive exists for developing a dry sodium SO₂ scrubbing systems. Using data from a recent study by Genco et al. (2) (with escalation to 1980) and a study by Burnett and O'Brien (3) indicates an installed capital cost of $122/kW for the dry sodium system compared to estimated costs of $186/kW for a wet limestone scrubbing system and $132/kW for a lime spray dryer system. These investment costs include SO₂ and particulate removal and waste disposal. The estimated operating costs for the dry injection SO₂ scrubber using nahcolite are 8.11 mills/kWh and 8.12 mills/kWh for refined trona, compared to an estimated 11.71 mills/kWh for the wet limestone scrubbing system and 8.55 mills/kWh for the lime spray dryer system. Figure 1 schematically presents the key features of a system that would inject sodium compounds into the flue gas after the preheaters; details on the wet limestone and dry lime flue gas desulfurization systems noted there are presented elsewhere.(3)

D.A. Furlong is now associated with ETS, Inc., Roanoke, Virginia.

SODIUM COMPOUND SUPPLY

Nahcolite ore is a naturally occurring mineral containing 70 to 90 percent sodium bicarbonate. It is found almost exclusively associated with oil shale. Vast resources of oil shale and associated nahcolite exist in the Eocene Green River formation in the Piceance Creek Basin of northwest Colorado. Sccording to the Univer States Bureau of Mines (4), this area is conservatively estimated to contain 26 billion metric tons* (29 billion short tons) of nahcolite. This amount of nahcolite could be used to desulfurize 533 billion metric tons (610 billion short tons) of 0.7 percent sulfur coal, assuming the requirement of 7.5 units of nahcolite per unit weight of sulfur removed. The adequacy of the nahcolite resource can be seen since the entire reserve of western bituminous and subbituminous coal is estimated at 390 billion metric tons (430 billion short tons).

In 1976, the Bureau of Mines launched a multi-year oil shale research and testing program to identify and resolve enviornmental problems associated with the development and mining of the deep deposits of oil shale and associated saline materials. A 3.05-m (10-ft) diameter pilot mine shaft was completed in 1978 at the Bureau's research facilities in Horse Draw, Rio Blanco County, Colorado. Nahcolite from this mine was provided for the testing program herein described. Efforts to develop a nahcolite mine are progressing** but are not yet a reality; hence, other sodium compounds are also being investigated.

Trona contains about 50 percent sodium carbonates. Major trona deposits are in the Green River formation in southwestern Wyoming. The total reserves in this area are estimated at 77 billion metric tons (85 billion short tons). The Green River trona is currently a major source of ore for commercial production of soda ash. Unfortunately, tax considerations (depletion allowances) do not favor the use of small quantities of trona as a raw ore.

Owens Lake in California is a source of

---

*1 metric ton = 1000 kg

**Multiminerals Corporation, a subsidiary of the Charter Company, is currently obtaining baseline nahcolite data for environmental impact purposes. Annual commercial production at a minimum of 1 million tons of nahcolite could be available in about 4 years.

sodium compounds available for immediate use. At least 45 million metric tons (50 million short tons) of sodium carbonates are available at Owens Lake. The raw ore from Owens Lake, upgraded only by mining methods, appears attractive for use as a sorbent in SO_2 removal. However, it also appears attractive to upgrade the raw ore to relatively pure sodium bicarbonate. A 1979 estimate indicates that 92 percent pure sodium bicarbonate could be produced at Owens Lake for approximately $55/metric ton ($50/short ton).(5)

Based on these considerations, three sodium compounds were selected for testing: (1) nahcolite, (2) crude ore (trona) from Owens Lake, and (3) sodium bicarbonate made from the Owens Lake ore by a carbonation and crystallization process. The nahcolite used (from the Bureau of Mines' Horse Draw mine) contained from 50 to 55 percent sodium bicarbonate as contrasted to the 70 to 90 percent indicated earlier in this section as naturally occurring.

TESTING

Testing of SO_2 removal by injection of dry sodium compounds has been carried out on a nominal 1.42-m^3/s (3000-cfm) slipstream on the Martin Drake Steam Plant Unit No. 6 of the City of Colorado Springs. This unit fires Colorado coals having sulfur contents of 0.4 to 0.6 percent. A schematic of the test setup is shown in Figure 1.

The sodium compounds were ground to 95 percent minus 200 mesh and then fed by a vibration screw feeder into a star wheel feeder for pneumatic conveying into the flue gas slipstream. Injection was countercurrent into the slipstream to increase turbulent mixing. The slipstream was an insulated, 0.38-m (15-in.) line, approximately 30.5 m (100 ft) long, leading to the 16-gag test filter. Fore-three tests were conducted for three sodium compounds with temperature, stoichiometric ratio, and air-to-cloth ratio as variables for the ranges shown in Table 1. Table 2 shows representative compositions of the sodium compounds. Instrumentation used is listed in Table 3. Test results are presented in Figures 2, 3, and 4 for nahcolite, trona, and refined trona, respectively. Since these results for a small-scale steady flowing slipstream served as the bases for the costs of the theoretical 500 MW plant in Table 4, these costs should be used with caution: slipstream conditions may not closely follow boiler cycle dynamics.

DISCUSSION OF RESULTS

The pilot scale project described in this paper sought to investigate three sodium compounds as potentially effective SO_2/NO_x suppressants through a series of 43 parametric tests covering a range of stoichiometric ratios, flue gas temperatures at the sorbent injection point, fabric filter compartment (baghouse) temperatures, and air-to-cloth ratios. The results of this work showed that all three compounds were effective at removing SO_2 from the slipstream flue gas.

Nahcolite was the best performer, yielding approximately 67 percent SO_2 removal at stoichiometric ratio (SR) = 1.0. Baghouse temperatures was 162°C (325°F), and injection was into the duct approximately 30.5 meters (100 feet) upstream. Increasing the baghouse temperature to 260°C (500°F) decreased the amount of removal to 38 percent at SR = 1.0. For continuous injection over a 30-minute test, an SR of 1.05 was required to achieve 70 percent removal while increasing to an SR of 2.0 yielded 91 percent SO_2 removal.

Trona was also capable of reducing SO_2, although removal rates were considerably lower than for nahcolite. At an SR of 1.0, trona yielded an SO_2 reduction of 23 percent from the same upstream injection location and at the 163°C (325°F) baghouse temperature, Increasing the temperature to 260°C (500°F) increased the SO_2 removal to 32 percent at SR = 1.0. Raising the SR to 1.6 improved removal efficiency to approximately 50 percent. No further increase in SO_2 removal occurred as the stoichiometric ratio was raised from 1.6 to 2.75.

Refined trona, a bicarbonate-enriched form of the same ore, performed much better than trona. At an SR = 1.0, it demonstrated 45 percent removal at the 163°C (325°F) baghouse temperature. Efficiency was slightly less (same trend as nahcolite) when the temperature was elevated to 260°C (500°F), showing 41 percent removal at SR = 1.0. An SR of 1.6 was required to achieve 70 percent removal. It yielded a maximum efficiency of 79 percent at SR = 1.75.

All three sorbents showed a slight (<10 percent) ability for removing NO_x. Also, NO_x removal was better at the lower temperatures.

Varying the air-to-cloth ratio from 1.44 to 2.99 had no noticeable effect on SO_2/NO_x reduction.

The nahcolite and refined trona are thought to be more effective at SO_2 adsorption (when compared to trona) because of their sodium bicarbonate composition. In addition the mineral matrix entrapping the sodium alkali may have enhanced SO_2 removal with nahcolite, relative to refined trona.

The stoichiometric ratio and feed rate were determined by total sodium purity. However, the bicarbonate form is considered to be significantly more reactive; thus for "total sodium parts" present, the reagent with the highest percentage of "bicarbonate parts" would be expected to perform the best. This expectation is supported by the higher SO_2 removal with nahcolite than refined trona as the sodium bicarbonate weight fraction of the total sodium carbonates in nahcolite exceeded that for refined trona. Similarly, the same trend was noted for refined trona relative to trona.

The percentage SO_2 in the flue gas when calculated as the product of coal combustion (using ultimate coal analyses) varied between 325 and 440 ppm. The calculated range in SO_2 concentration was consistent with the measured inlet SO_2 concentrations in the slipstream.

SYSTEM COMPARISON

SO_2 removal by dry sodium compounds has been compared with SO_2 removal by wet scrubbing with limestone and by spray absorption using lime. Of course, to make such a comparison, many assumptions need to be made. This is difficult to do fairly without making a study of a specific site. However, an attempt was made to make such a comparison at least to show generally the strengths and weaknesses of each system. Table 4 lists the results of the study along with reported costs for wet limestone and spray drying SO_2 removal systems. Table 4 indicates that the dry scrubbing systems have much lower capital and annual costs than wet scrubbing and that dry injection using sodium compounds is slightly cheaper than spray drying. Dry scrubbing with nahcolite and refined trona gave the same capital and essentially the same annual revenue requirements. Dry injection capital and annual revenue requirement costs were less than 10 percent lower than those for spray drying, which is minor considering the -40 to +10 percent accuracy for these preliminary cost estimates. However, the dry injection system is simpler and would be expected to be more reliable than the spray dryer.

It is emphasized that wet limestone SO_2 removal systems are operational while to date only pilot plant data are available for the two dry flue gas desulfurization options listed in Table 4. Although spray dryers have been marketed commercially and two units for industrial boilers were started up in 1979, performance and cost data for these operational units are not yet available.

WASTE DISPOSAL

Sodium salts resulting from the removal of SO_2 from flue gases with sodium compounds are readily soluble in water. Consequently, disposal of these wastes from dry SO_2 removal processes is a major environmental concern. Concurrent with the present work, options for disposing of scrubber wastes are being studied using wastes from the dry injection tests at Colorado Springs. Under subcontract from Buell-Envirotech, Battelle Memorial Institute (Columbus Laboratories) is assessing waste disposal via insolubilization of wastes before disposal or reuse.

SUMMARY

Pilot tests indicate that dry injection of sodium compounds into flue gases from a western-coal-fired boiler, followed by particulate collection in fabric filters, is a technically feasible and an economically attractive approach to flue gas desulfurization. This type of dry SO_2 removal appears to have both capital and operating cost advantages over spray drying and wet limestone processes as well as process simplicity. The estimated supply of nahcolite alone for dry SO_2 removal is more than sufficient for the entire reserve of western bituminous and subbituminous coal. Potential additional sources of sodium compounds for dry SO_2 removal include trona and Owens Lake ore; the latter is available for immediate use, while tax considerations do not presently favor the use of trona. Wastes resulting from using sodium compounds for removing SO_2 from flue gases are a problem which is being addressed in small-scale tests now underway.

ACKNOWLEDGMENTS

The authors acknowledge with appreciation the contributions of John Urich and Dennis Lapp of Buell-Envirotech in collecting the dry injection data reported here. The assistance of Dennis Lapp in data reduction and analysis is also gratefully acknowledged. This work was supported under EPA Contract No. 68-02-3119 and with the cooperation of the City of Colorado Springs.

LITERATURE CITED

1. Blythe, G.M., J.C. Dickerman, and M.E. Kelly, "Survey of Dry SO_2 Control Systems", EPA-600/7-80-030 (NTIS PB80-166853), U.S. Environmental Protection Agency, Washington, DC (1980).

2. Genco, J.M., H.S. Rosenberg, M.Y. Anastas, E.C. Rosar, and J.M. Dulin, J. APCA 25(12):1244 (1975).

3. Burnett, T.A. and W.E. O'Brien, "Preliminary Economic Analysis of a Lime Spray Dryer FGD System", EPA-600/7-80-050 (NTIS PB80-190051), U.S. Environmental Protection Agency, Washington, DC (1980).

4. Beard, T.N., D.B. Tait, and J.W. Smith, "1974 Nahcolite and Dawsonite Resources in the Green River Formation, Piceance Creek Basin, Colorado", Rocky Mountain Association of Geologists, 1974 Guidebook, pp 101-109.

5. McClung, C.W., Private Communication to Dale A. Furlong, Buell Emission Control Division, Envirotech Corporation, Lebanon, PA, (1979).

Table 1. Test Variables and Their Ranges

Variable	Range
Sodium Compound	Nahcolite, Trona, Refined Trona
Injection Temperature	204 to 327°C (400 to 620°F)
Baghouse Temperature	163 to 260°C (325 to 500°F)
Stoichiometric Ratio	0.7 to 2.1
Air-to-Cloth Ratio	0.46 to 0.91 m/s (1.5 to 3.0 ft/min)

Table 2. Chemical Composition of Sodium Compounds

Test Run	Nahcolite		Refined Trona	Trona
	N-5A	N-5B	B-1	T-1
Constituent				
$NaHCO_3$	52.76	54.01	59.32	26.43
Na_2CO_3	3.64	3.13	14.22	41.39
NaCl	0.49	0.46	1.27	4.14
Na_2SO_4	0.40	0.32	1.19	5.19
Na_2SO_3	0.001	0.002	<0.001	<0.001
$NaNO_3$	0.02	0.03	<0.01	<0.01
Insolubles	40.86	40.01	15.34	9.72
Water	1.83	2.04	8.65	13.12

Table 3. Test Instrumentation

Parameter	Instrument
SO_2	TECO Model 40 Pulsed Fluorescent
NO/NO_x	TECO Model 10 Chemiluminescent
O_2	Teledyne Model 9500X Electro-Chemical
Sodium Compound Flow	Grab Sample, Timed Weight
Flue Gas Flow	ASME Venturi
Temperatures	Thermocouples (Type J)

Table 4. Estimated Costs of SO$_2$ Removal Systems

| | Wet Scrubbing | Dry Scrubbing | | |
| | Venturi/Spray Tower | Spray Dryer | Dry Injection | |
	Limestone	Lime	Nahcolite	Refined Trona[c]
Sorbent				
Cost, $/metric ton	9.4	112.4	41.8	23.7
($/short ton)	(8.5)	(102.0)	(37.9)	(21.5)
Stoichiometric Ratio	0.90	1.0	1.05	1.6
Requirement, kg/kW-yr	48.1	30.9	121.6	184.2
(lb/kW-yr)	(106.0)	(68.1)	(268.0)	(406.0)
Capital Cost[a], $/kW	186.38	132.3	121.6	121.6
Annual Revenue Requirement[a], mills/kWh	11.71	8.55	8.11	8.12

a. Based on 1984 dollars for plant described in b.

b. Includes SO$_2$ removal, particulate collection, and waste disposal on plant site.[3]
    Assumes compliance with 1979 New Source Performance Standards for Utility Boilers.
    Other premises:

Item	Premise
Power plant	New, Great Plains – Rocky Mountain region, 500 MW coal-fired boiler, 10.1 MJ/kWh (9,500 Btu/kWh) heat rate
Operating schedule	130,400 hr, 30-yr life, 5,500-hr first-year operation
Fuel	Subbituminous coal; 22.7 MJ/kg (9,700 Btu/lb), 0.7% sulfur, 9.7% ash, 16% moisture; Colorado coal (0.4-0.6% sulfur) for dry injection
Base year	Capital investment: mid-1982
	Revenue requirements: 1984
FGD waste disposal	Wet scrubbing: clay-lined pond
	Dry scrubbing: landfill
SO$_2$ removal efficiency	70%
Particulate removal efficiency	99.8% (13 ng of particulates/kJ heat input (0.03 lb/MBtu))
SO$_2$ absorber redundancy	33% (3 operating trains, 1 spare) for venturi/spray tower and spray dryer; none for dry injection

c. Trona, as mined, did not give 70% removal at stoichiometric ratios tested (up to 2.75, see Figure 3).
    Refined trona, a sodium-bicarbonate-enriched trona, gave 70% removal at a stoichiometric ratio of 1.6
    (see Figure 4).

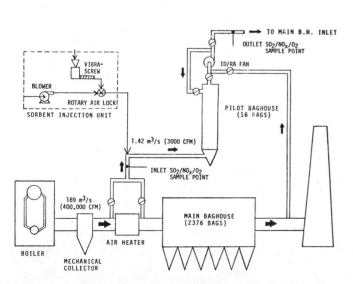

Figure 1. Dry sodium injection/fabric filter SO$_2$ removal system on slipstream at City of Colorado Springs, Martin Drake Unit No. 6.

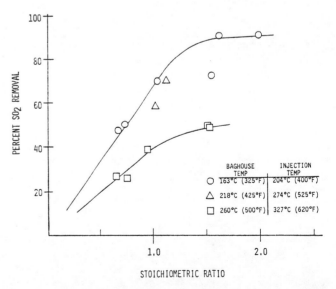

Figure 2. SO$_2$ removal with dry injection of nahcolite into duct of pilot system.

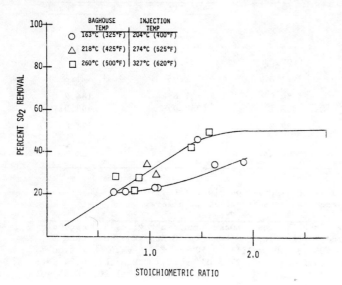

Figure 3. SO₂ removal with dry injection of trona into duct of pilot system.

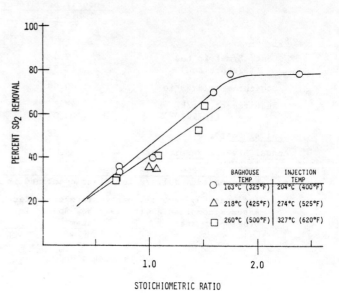

Figure 4. SO₂ removal with dry injection of refined trona into duct of pilot system.

CHEMICAL CHARACTERIZATION OF ATMOSPHERIC AEROSOL USING ATTENUATED TOTAL INTERNAL REFLECTION (ATR) INFRARED SPECTROSCOPY

STANLEY A. JOHNSON
PAUL T. CUNNINGHAM
and
ROMESH KUMAR

Chemical Engineering Division
Argonne National Laboratory
9700 South Cass Avenue
Argonne, Illinois 60439

During the last decade, a wide variety of techniques for the collection and analysis of airborne particulate matter has been applied to the chemical characterization of atmospheric aerosol. In this paper, we discuss the results obtained with a new instrument which combines several conventional techniques in a unique manner, and which has the potential to supply sensitive, real-time analysis of the sulfate, nitrate, and acidity in airborne particles. The instrument uses inertial impaction for sample collection, and infrared spectroscopy for analysis, with attenuated, multiple total internal reflection providing enhancement of the absorption spectra. The ATR-Impactor instrument (for Attenuated Total internal Reflection-Impactor) has been described elsewhere (1). Briefly, the aerosol to be analyzed is drawn through a cascade virtual impactor to remove particles greater than $\sim$1.2 μm aerodynamic diameter. Particles between $\sim$0.5 and 1.2 μm are then deposited on the opposite surfaces of an internal reflection element (IRE). A set of IRE's moves past the impaction nozzles to provide sample time resolution. Figures 1 and 2 schematically show the design of the ATR-Impactor. In the present version of the instrument, the material deposited on the IRE's is analyzed using Fourier-transform infrared spectroscopy (FT-IR). The principles of inertial impaction for sampling (see Ref. 2, for example) and infrared spectroscopy for analysis (3 to 5) are well known. Attenuated total internal reflection theory has been discussed by Harrick (6). At each internal reflection, the light beam is attenuated by light absorbing species in particles which are within about one-half wavelength of, and external to, the surface. Due to the multiple internal reflections, ATR is a very sensitive technique, yielding an absorption spectrum which closely resembles that obtained by conventional methods. Subsequently, the collected sample is available if analysis by other means is desired.

RESULTS AND DISCUSSION

In the first field trials of the ATR-Impactor, results from the ATR method and the conventional KBr pellet method were compared. Figure 3 shows the infrared spectra of ambient samples collected simultaneously and analyzed using the two methods. As expected for the submicrometer-sized ambient particulate matter, the spectra disclose the presence of NH_4^+ (1400 cm^{-1}), SO_4^{2-} (620 and 1110 cm^{-1}) and NO_3^- (1384 and 840 cm^{-1}) ions. However, as can be seen in Figure 3, the sulfate absorbance at 620 cm^{-1} is at least six times greater in the ATR spectrum than in the KBr pellet spectrum of the same amount of sulfate. Similarly, the 1110 cm^{-1} sulfate band, the 1400 cm^{-1} ammonium band, and the 840 cm^{-1} nitrate band are significantly enhanced in the ATR spectrum. It should be mentioned that all the bands are slightly shifted between the two types of spectra. Laboratory experiments with known materials have verified the positions of these shifted bands.

0065-8812-81-3815-0211-$2.00

Some general observations have been made during the initial work with the ATR-Impactor. Microscopic examination of the sample on the IRE shows that, for a one-hour collection at 30 l/min, where ~10 µg of sulfate were collected, particulate matter covered less than 25% of the IRE surface. This low coverage is significant since interaction between collected particles is minimized. Spectra have been obtained from ambient samples with 15-minute time resolution (15 minutes per IRE) where the IR absorption bands were well above noise level and easily identifiable. Coverage of the IRE surface was observed to be less than 10% in these cases. The limit of detection is estimated to be well below 0.5 µg of sulfate per IRE. In moderate loading of the air, time resolution of five minutes should be easily attainable with the present configuration of the instrument.

The instrument was initially designed for use under ambient conditions. To extend the applicability to airborne power plant plume studies, a set of nozzles has been fabricated with appropriate dimensions to allow a flow rate of 60 l/min. This doubling of the flow rate will provide time resolution of less than one minute/IRE and permit the analyses of samples collected during the rather short transit time through the plume.

Perhaps the most challenging aspect of the ATR-Impactor is its potential for real-time chemical characterization of atmospheric aerosol. The coupling of FT-IR directly with sample collection through the use of ATR has greatly reduced sample handling and preparation, and has thereby shortened the time between sample collection and analysis. However, even under ideal conditions, some time elapses while the IRE is removed from the impactor and placed in the spectrophotometer. Incorporation of spectral measurement capability directly into the ATR-Impactor instrument would provide analysis as the sample is being collected.

SUMMARY

The ATR-Impactor has successfully completed initial laboratory and field tests and has shown that it offers several important advantages for the spectroscopic chemical characterization of ambient aerosol. It has been shown that the absorbance of the infrared bands of the collected material is at least six times more intense than for the conventional KBr pellet technique. For sulfate, the limit of detection is thus less than 0.5 µg. This is sufficient for five-minute time

resolution in the analysis of typical ambient concentrations of sulfate using the prototype unit with a flow rate of 30 l/min. To make the ATR-Impactor applicable for plume studies the air flow rate has been doubled to 60 l/min to attain a time resolution of less than one minute/IRE which is needed for aircraft flights through plumes.

Other advantages lie in the virtual elimination of the time-consuming sampling handling and preparation procedures. This makes feasible the incorporation of spectral measurement devices directly into the sampling unit, thereby providing for analysis as the sample is being collected.

ACKNOWLEDGEMENT

This work was performed under the auspices of the U.S. Department of Energy.

LITERATURE CITED

1. S. A. Johnson and P. T. Cunningham, American Chemical Society, Division of Environmental Chemistry, Preprints of Papers Presented at the 177th National Meeting, 19 (1), Paper Number 266 (1979).

2. V. A. Marple and K. Willeke, Fine Particle Symposium, Minneapolis, Minnesota, May 28-30, 1975, Particle Technology Laboratory Publication No. 251.

3. P. T. Cunningham, S. A. Johnson and R. T. Yang, Environ. Sci. Technol. 8, 131 (1974).

4. P. T. Cunningham and S. A. Johnson, Science 191, 77-79 (1976).

5. R. Kellner and T. Novakov, Lawrence Berkeley Laboratory, Energy and Environmental Division Annual Report 1978, LBL-8619, 189-192 (1978).

6. N. J. Harrick, Internal Reflection Spectroscopy, Interscience Publishers, New York (1967).

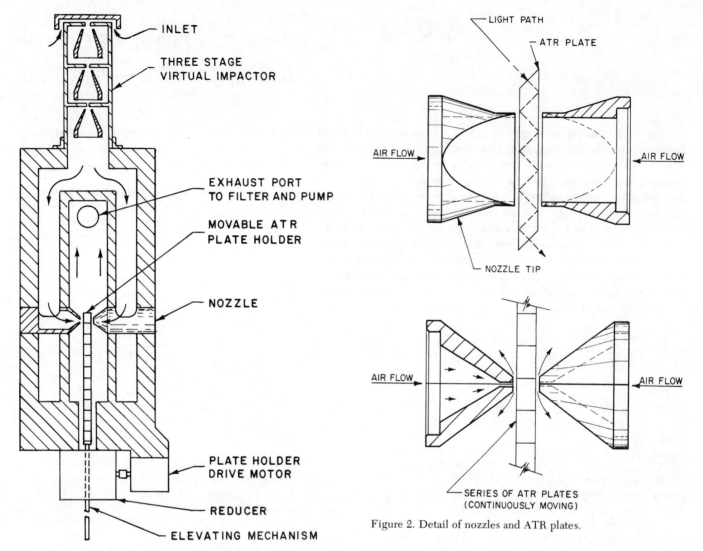

Figure 1. Schematic diagram of ATR-impactor.

Figure 2. Detail of nozzles and ATR plates.

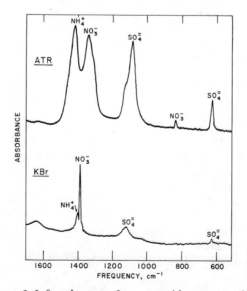

Figure 3. Infrared spectra for comparable amounts of airborne
particulate matter from KBr pellets and ATR.

COAL DESULFURIZATION BY CHLORINOLYSIS—PHASE II

JOHN J. KALVINSKAS

and

NARESH ROHATGI

Jet Propulsion Laboratory
California Institute of Technology
Pasadena, California

Coal desulfurization by low temperature chlorinolysis was conducted by the Jet Propulsion Laboratory, California Institute of Technology. An engineering scale batch reactor system was constructed and operated for the evaluation of five high sulfur bituminous coals obtained from Kentucky, Ohio, and Illinois. Forty-four test runs were conducted on 2 kilograms of coal per batch under conditions of 100×200 mesh coal, solvents—methylchloroform and water, 333-403°K, 101-514 kPa, 45 to 90 minutes reaction time and gaseous chlorine flow rates of up to 0.67 sm³/h. Experimental data are presented for sulfur forms, proximate, ultimate, ash elemental and trace element analysis on the raw and treated coals along with coal filtrate and wash water analysis. In addition, an integrated continuous flow mini-pilot-plant was designed and constructed for a nominal coal feed rate of 2 kilograms per hour which will be operated as part of the Phase III program. A equipment flow sheet of the mini-pilot-plant is presented.

EQUIPMENT

The coal desulfurization experimental work was carried out in a bench-scale batch reactor system, depicted in Figure 1. The batch reactor, Figure 2, was constructed of 45.7 cm (18-inch), schedule 40, mild steel pipe lined with 0.64 cm (1/4-inch) triflex semi-hard rubber membrane and three layers of red-shale type "L" acid-proof brick laid with 0.32 cm (1/8-inch) joints using Pennwalt's asplit CN mortar. The acid-resistant brick construction was necessary to accommodate the highly corrosive and erosive nature of the chlorinated coal slurry containing Cl_2, HCl, H_2SO_4 and coal suspended in either methylchloroform or water. The reactor cavity is 17.8 cm (7-inches) I.D. by 58.4 cm (23-inches) deep with provisions for 2 kilograms of coal suspended in 4 kilograms of solvent. The reactor was designed for operating up to 423°K (150°C) and 791 kPa (100 psig). The coal slurry was kept in suspension by a 186 watt (1/4 H.P.) Chemineer agitator with four 45 degree pitch turbine blades, 2.54 cm (1-inch) by 11.4 cm (4-1/2 inch) diameter operating at shaft speeds from 40 to 565 r.p.m. The reactor was equipped with a rupture disc, aluminum coated with Teflon, for a burst pressure of 1135 kPa (150 psig). Gaseous chlorine injection was from a standard liquid chlorine storage cylinder equipped with hot water immersion heating. A Linde flow-meter monitored chlorine flow over the range of 0.01 to 1.7 sm³/h (0.42 to 61.5 SCFH). Chlorine dispersion was through Teflon tubing drilled with 0.79 mm (1/32-inch) diameter holes and located to the side and near the reactor bottom. Pressure and temperatures were monitored continuously on a Hewlett Packard recorder and a Northrup Speedomax multi-point recorder. A Teflon covered stainless steel cooling coil in the reactor and direct steam injection provided temperature control. A 3.8 cm (1-1/2-inch) diameter ball valve connected the reactor to a 316 stainless steel reflux condenser followed by a 316 stainless steel $0.96 \times 10^{-2}m^3$ (0.34 cubic feet) capacity gas holder for containing and sampling off-gases from the reactor. A Fisher type 9811 pressure relief valve provided control from 205 to 791 kPa (15 to 100 psig).

A batch vacuum filtration unit fabricated from 2:1 elliptical tank heads 45.7 cm (18-inch) diameter, 12 gauge of 304 stainless steel with a filtration area of 0.17 m² (1.8 square feet) accommodated the filtration and wash of two kilograms of coal. A 10 mesh, 304 stainless steel wire screen supported a 325 mesh, 304 stainless steel wire filtration screen. An exhaust fan provided 50-75 cm (20-30 inches) water column vacuum for filtration. Wash and filtration water were recovered in the filtrate tank.

0065-8812-81-4527-0211-$2.00

The dechlorination unit, Figure 3, was designed to dechlorinate the chlorinated coal at atmospheric pressure and temperatures of 673 to 773°K (400 to 500°C). A Lindberg Model 58331 tube furnace was used in conjunction with a 304 stainless steel, 12.7 cm (5-inch) diameter by 1.5 m (5-foot) long, 0.64 cm (1/4 inch) wall tube. The furnace had three zone temperature control for up to 50 amps, 240 VAC, single phase, 50-60 Hz power designed to operate from 373 to 1373°K (100 to 1100°C). A variable speed drive for tube rotation at 1 to 20 r.p.m. was installed. Nitrogen purge gas was generally set at 0.85 sm^3/h (30 SCFH) and 105 kPa (0.5 psig). Temperature control was within 5°K (5°C) and provided by a sheafed iron-constantan thermocouple installed from one end and in direct contact with the coal bed near the bottom and longitudinal center. Corrosion did not pose a problem in the dechlorinator despite the presence of HCl and SO$_2$ off gases.

OPERATING PROCEDURE

Chlorination

The reactor was operated with two kilograms of coal and four kilograms of solvent kept in suspension by agitation at 565 r.p.m. with gaseous chlorine injected at a nominal flow rate of 0.28 Sm^3/h (10 SCFH). The cooling coil and direct steam injection provided temperature control. Coal slurry samples of 100 grams were obtained at 15, 30, 45 and sometimes 90 minutes. Sampling was near the wall and reactor bottom with stirring sufficiently intense to insure a representative coal slurry sample. Rapid reactor pressurization was obtained at high chlorine flowrates, greater than 1.56 Sm^3/h (>20 SCFH) whereas reduced chlorine flowrates, less than 0.28 Sm^3/h (<10 SCFH) were readily absorbed into the coal slurry until the slurry was saturated with chlorine, i.e. 45 minute times at 0.28 Sm^3/h (10 SCFH) chlorine flow. Reactor operation was normally with the reflux condenser closed from the reactor but a few runs provided a continuous vent of chlorine through the reactor and to the reflux condenser.

With methylchloroform solvent runs, four kilograms of water were added to the reactor and live steam was injected to strip the solvent from the reactor to the reflux condenser and solvent recovery tank. With water as the suspension medium, the coal slurry was allowed to cool and then drained into a holding tank for coal slurry filtra-

tion and water wash on the batch vacuum filter.

Vacuum Filtration - Spray Wash

The coal slurry was filtered and then spray washed for a displacement of the mother liquor at a water/coal addition of 2/1. Wash temperatures were initially at 353 to 373°K (80 to 100°C) and then reduced to 298°K (25°C) without any effect. Adequacy of the water wash was monitored by the residual sulfate determination in the washed coal.

DECHLORINATION

Coal removed from the filtration unit was dried in a vacuum oven at 373°K (100°C) overnight and then stored in closed glass containers for up to three months before being dechlorinated.

Dechlorination was carried out by charging the total amount of coal, up to 2 kilograms, into the dechlorinator which had been preheated to 673°K (400°C). Approximately thirty minutes were required to heat the coal to 673°K (400°C) while maintaining a nitrogen purge of 0.84 Sm^3/h (30 SCFH). The coal was then held at 673°K (400°C) for 30 to 60 minutes. Tube rotation was maintained at 4 r.p.m. After 30 to 60 minutes of cooling the dechlorinator, the coal was removed and stored in a closed glass container.

Analyses

Coal analyses conducted were for sulfur forms, ultimate, proximate and trace elements of the raw and processed coals. The Colorado School of Mines Research Institute was used for conducting the coal analyses. The Eschka method was used for total sulfur and ASTM procedures were used for pyritic and sulfate sulfur with organic sulfur determined by difference.

Comparative analyses were also conducted of given samples by: Galbraith Laboratories, Knoxville, Tennessee; Standard Laboratories, Charleston, West Virginia and the DOE coal analysis laboratory at the Pittsburgh Mining Technology Center. The majority of analyses conducted by the different laboratories were in substantial agreement.

For a rapid determination of total sulfur in the processed coal samples, JPL used a Leco acid-based analyzer. Care was taken to dechlorinate the coal samples before Leco analyses. This analysis was used only as a preliminary estimate of total sulfur content.

Wash and filtrate solutions were analyzed for sulfates, chlorides, iron and trace elements to complete material balances.

Mass spectroscopy analyses were conducted on gas samples from the chlorinator for runs with methylchloroform and water.

COALS

Selection

The five (5) coals selected for the bench-scale batch reactor tests are listed in Table 1 with attendant analyses for sulfur forms, proximate analyses, and ultimate analyses. They are bituminous coals obtained from Ohio, Illinois and Kentucky and were selected for testing by JPL and the DOE Pittsburgh Mining Technology Center. Forty five hundred kilograms each of PSOC 276 and PSOC 282 coals were obtained directly from the Georgetown No. 24 and Orient No. 6 mines, respectively. Approximately 45 kilograms of Island Creek Coal were obtained from the Pittsburgh Mining Technology Center. The remaining coals PSOC 219 and PSOC 026 were obtained from the Penn State Coal bank and are representative of the coals tested in the Phase I laboratory test program.

Raw Coal Analyses

All of the analyses of the raw coals listed were obtained from the Colorado School of Mines Research Institute. In most instances, the analyses were substantially different from the Penn State Coal bank analyses for these coals.

Sulfur forms analyses were generally conducted for raw coal samples used in each test run. The standard deviation was calculated for each of the sulfur forms analyses for each of the coals and represents the discrepancy between runs of sulfur forms that can be attributed to a combination of non-uniformity in coal samples and/or analytical deviations. For PSOC 276 coal with eleven samples, standard deviations were: organic sulfur ±14%, pyritic sulfur ±8% and total sulfur ±2%. Average values of sulfur forms and standard deviations are listed for each of the coals in Table 1. PSOC 282 coal was analyzed for two mesh sizes, -100 to +200 mesh and -16 to +100 mesh. Sulfur forms were

not significantly different as a result of this particle size difference.

Proximate analysis, ultimate analysis, ash elemental analyses, and trace elements were determined for three (3) coal samples each for PSOC 276 and PSOC 282 and single coal samples for PSOC 210, 026 and Island Creek coal. Analysis and standard deviations are listed in Table 1. Trace elements analyses include Pb, As, Se, Hg, V.

Size Distribution

Eighteen hundred kilograms each of coals PSOC 276 and 282 were ground from run of the mine coal and then sieved in a trommel screen to provide coal samples for testing in the range of -100 to +200 mesh size and also in the range of -16 to +100 mesh. Analyses of both sizes of coal samples were conducted to obtain the size distribution as shown in Table 2. For -100 to +200 mesh only 0.7 wt. % was larger than +100 mesh and 16.9 wt. % was finer than 200 mesh. For -16 to +100 mesh, 7 wt. % was greater than 16 mesh, with approximately 47.8 wt. % of the coal between 16 and 60 mesh and 20.2 wt. % finer than 100 mesh.

EXPERIMENTAL DATA

A total of 44 test runs (2) were conducted with 15 runs on Coal PSOC 276, 19 runs on coal PSOC 282, 2 runs on PSOC 219, 3 runs on PSOC 026 and 5 runs on Island Creek coal (Western Kentucky Union County No. 9 Seam). A summary of operating conditions and desulfurization data are presented, Table 3, for the 5 bituminous coals. The data are grouped by coal, methylchloroform and water solvent, reaction times of 15, 30, 45, 90 minutes, bulk and dechlorinated coal. The samples for the given reaction times are grab samples obtained from the reactor. The bulk sample is a sample of the coal after it has been fully processed including solvent separation by distillation, coal slurry filtration and spray wash. The dechlorinated sample is that obtained after the thermal dechlorination of the bulk coal has been accomplished in the Lindberg furnace. Additionally, the data for PSOC 282 has been segregated by mesh size with the nominal size at 100 × 200 mesh and a coarse mesh of -16 × 100 mesh. In the case of both PSOC 282 coal and Island Creek coal, staging of the coal desulfurization was made in two and three stages with the processed coal from the first stage being filtered and washed and used as the second stage charge with fresh methylchloroform or water being added.

Inspection of the individual runs with respect to temperature, pressure and chlorine flow rate for the respective coals did not show any noticeable correlation with the sulfur forms data over the range of operating conditions. Operating conditions were in the range of 333-403°K, 101-514 kPa, chlorine feed rate of 0.14 to 0.67 Sm^3/h. Methylchloroform runs were generally confined to 338°K with a few runs up to 373°K and water runs were conducted at 338 to 403°K. Changes in gaseous chlorine injection from a 6.3 mm tube opening located to the side and beneath the agitator impeller to a standard fritted glass diffuser element and finally to a Teflon tube drilled with 0.34- to 6.3-mm diameter holes with a final size of 3.2-mm holes did materially affect the introduction of chlorine into the coal slurry solution but did not translate into a corresponding effect on coal desulfurization.

A substantial variation in temperature, pressure and chlorine flow rates existed between runs so that any substantial effect of these variables on the coal desulfurization reaction would have been evident if a large effect existed. Reaction times of 15 and 30 minutes were sufficiently short to allow observation of kinetic reaction effects in this operating range. A reaction time of 45 minutes provided a leveling off and/or peaking of coal desulfurization.

Sulfur Forms

The average of sulfur forms reductions are indicated for groups of runs listed by coal, solvent and reaction time, Table 3, for the range of operating conditions designated. Two and three staged reaction run data are listed individually along with individual run data with a coarse (16-100 mesh) PSOC 282 coal.

PSOC 276. This coal shows a negative organic sulfur removal, a higher pyritic sulfur removal and higher total sulfur removal with methylchloroform than with water and a significant effect of dechlorination on increasing the desulfurization by about 10% (3-5 absolute %). Peak total desulfurization is 60% with methylchloroform and 43% with water after dechlorination. In the remaining 4 coals tested, methylchloroform relative to water assisted organic sulfur removal but did not aid total sulfur removal. The negative organic sulfur removal reported is ascribed to analytical bias rather than the formation of organic sulfur compounds in the processed coal.

PSOC 282. This coal shows a high organic sulfur removal with methylchloroform (58% at 45 minutes and 82% at 90 minutes) with a sharply reduced value for bulk and dechlorinated samples. The anomaly in organic sulfur removal may be the result of an interference of methylchloroform with the pyritic sulfur determination and as a consequence with the organic sulfur determination (which is obtained by difference between total and pyritic sulfur values). Organic sulfur removal with water as the solvent does not show a high organic sulfur removal in the grab samples (15, 30, 45 minute samples) but the bulk and dechlorinated samples show comparable organic sulfur removal with methylchloroform and water. The relatively low pyritic sulfur removal with methylchloroform also suggests that the pyritic sulfur determination may be incorrect (at the high residual pyritic sulfur value). Dechlorination shows a 16 to 29% improvement in desulfurization of the bulk sample. Extended reaction times beyond 45 minutes to 90 minutes provides no increase in total desulfurization with methylchloroform but a significant improvement with water. The use of a coarse mesh, 16 × 100 mesh shows one-third less total desulfurization at 45 minutes and somewhat less, 8-19%, at 90 minutes. The two-stage desulfurization indicates a significant improvement in sulfur removal with the second stage over the first stage, approximately one-third improvement in total desulfurization.

PSOC 219. This coal showed a high organic sulfur removal with grab samples at 15, 30, 45, and 90 minutes (31, 37, 62% and 100% resp.) that is not exhibited in the bulk sample. The water solvent shows no comparable organic sulfur removal to that exhibited with methylchloroform. Total sulfur removal is comparable between methylchloroform and water solvents. Dechlorination provides a 14% increase in total desulfurization to 66%.

PSOC 026. This coal shows a high organic sulfur removal with methylchloroform in the grab samples but falls to a low value in the bulk sample, comparable to that obtained with the water solvent. Total sulfur removals are comparable between the methylchloroform and water solvents. The dechlorination stage provides an 18% improvement in total desulfurization.

Island Creek. This coal with methylchloroform shows a 21% increase in total desulfurization for the second stage over the first stage with a accompanying sharply improved organic sulfur removal, although again the grab samples indicated the anomaly of a higher organic sulfur removal than that found in the bulk sample. The three stage desulfurization with water shows a low total desulfurization for the first stage of 29% for the bulk sample and a 59% improvement in the second stage for total desulfurization but no increase in total sulfur removal for the third stage.

Proximate Analysis

Proximate analyses for the five raw and desulfurized coals are included in Table 4. Individual changes are included for each of the coals. Average changes for the 5 coals indicate a 29% reduction in volatile matter, a 24% reduction in ash, a 25% increase in fixed carbon (corresponds to the decrease in volatiles), and a 5.3% decrease in heating values.

Ultimate Analysis

Ultimate analyses for the 5 raw and desulfurized coals are included in Table 5 for the individual and combined coal values. The average for the 5 coals indicates a 3.7% increase in carbon, a 36.6% decrease in hydrogen, a 4.1% decrease in nitrogen, a 20% increase in oxygen, a 1067% increase in chlorine, and a 55% decrease in sulfur.

Ash Elemental Analysis

The ash composition for the raw and desulfurized coals is included in Table 6 for the five bituminous coals. Ash elements and average ash element reductions for the five coals are: Si (-16%), Fe (59%), Al (0%), Ca (71%), K (0%), S (55%), Ti (-14%), Na (29%), Mg (0%), P (0%), Mn (0%), total ash (20%).

Trace Element Analysis

Trace element analysis for the raw and desulfurized coals are listed by coal and trace elements, Table 7. Average trace element reductions for the five coals are: Lead (23%), Arsenic (40%), Selenium (-20%), Mercury (67%). Vanadium levels were below the analytical detection level of 50 ppm in the raw and desulfurized coals.

Material Accounting

The average batch reactor feed composition for the five bituminous coals is listed in Table 8. The average coal filtrate and wash water analysis for the 5 bituminous coals is listed by concentration (mg/ℓ) in the fil-trate and wash water and by (weight percent) of the dry product coal, Table 9. The principal components removed from the coal and in solution are Cl^-, SO_4^- and Fe^{+++}. Elements removed in trace quantities are Ca^{++}, Na^+, Al^{+++}, Mg^{++}, and K^+.

The average overall material accounting is listed in Table 10 for the five bituminous coals tested. The overall average accounting for the five coals is: raw coal - 100 ± 3%, coal organic fraction - 99 ± 3%, ash - 114%; sulfur - 94%, chlorine - 86%; methylchloroform - 83 ± 7%. The high ash accounting is attributed to the contribution from corrosion of stainless steel tubing, impeller, test coupons, etc. that were dissolved into the coal slurry. The material accounting includes the chlorination, distillation and coal filtration wash. A material balance across the dechlorination stage is separate from the above material accounting and follows.

Coal Dechlorination

Coal dechlorination data are summarized for the average of the 5 bituminous coals, Table 11. Dechlorination conditions were 60 to 95 minutes at 673°K with a nitrogen purge of 0.84 Sm^3/h over a average coal feed for the 5 bituminous coals to the dechlorinator of 1078 - 1372 grams, and a average feed coal chlorine composition of 6.26 - 25.0 weight percent. Chlorine removal for the five coals averaged 87.0 to 93.0%. Product coal recovery was an average 86.7 to 91.1%. Relatively large mechanical losses were observed in charging the dechlorinator with the coal and in the recovery of coal product. Subsequent improvements in coal handling to the dechlorinator have substantially minimized the losses.

CONTINUOUS FLOW-MINI-PILOT PLANT

The mini-pilot plant for coal desulfurization by low temperature chlorination is designed to feed pulverized coal (14-200 mesh) at a nominal rate of 2 kg/hr. Nominal solvent and water flow rates are 4 kg and 1.4 kg per hour respectively. The plant is designed to operate at a pressure and temperature ranging from 101 to 790 kPa and 323 to 423°K. The equipment flow schematic for the mini-pilot plant is shown in Figure 4. Nominal design variation in flow rates with corresponding changes of retention time are in the range of -50 to +100% variation. Stainless steel surfaces in the chlorinator will be protected by Teflon coating to overcome corrosion problems. The mini-pilot-plant was constructed as part of the Phase II program and will be operated as part of the follow-on Phase III program.

CONCLUSIONS

The bench scale batch reactor studies with 2 kilograms of coal per batch provided a broader range of pressure, temperature, and increased chlorine flow rates than that explored in the earlier laboratory scale studies. Unfortunately, the promise of improved coal desulfurization with increases in operating temperature, pressure, and chlorine flowrates did not materialize. Apparently, the chlorine concentration in solution does not control the desulfurization reaction kinetics under the range of operating conditions tested.

The introduction of water as a solvent in lieu of methylchloroform shows considerable promise for providing comparable total sulfur removal, although organic sulfur removal appears to be favored by the use of methylchloroform solvent.

Apparent anomalies between organic and pyritic sulfur values suggest a need to improve the determination of pyritic sulfur (and consequently organic sulfur determination). An apparent analytical bias may exist by relying on the iron (Fe) determination in the pyritic sulfur determination instead of using a direct sulfur determination obtained by the barium sulfate precipitate method.

Material balances on coal indicate a high (99%) recovery of the organic coal fraction from the chlorination through coal slurry filtration - wash stage of the desulfurization process. Relatively large coal losses (10%) were found in the dechlorination stage which were attributed primarily to observed mechanical handling losses.

Coal dechlorination requires added improvements in order to reduce chlorine levels to that found in the raw, unprocessed coal of 0.1 to 0.5 weight %. Initial lab scale dechlorination tests (Phase I) achieved chlorine levels of less than 0.1 weight percent. The dechlorination process also promotes a reduction in coal volatiles by increasing the fixed carbon values (cross-linking) with a 5% loss in heating value attributed to coal hydrogen loss as HCl in dechlorination.

A substantial average ash reduction of 20 to 25% for the five bituminous coals was found for the desulfurization process.

FUTURE WORK

Additional batch reactor testing on low and high sulfur bituminous, sub-bituminous, and lignite coals is planned under the Phase III program to provide improvements in coal desulfurization and an improved understanding of the process.

The continuous flow mini-pilot-plant will be operated as part of the Phase III program.

ACKNOWLEDGEMENT

The work reported herein is part of the Coal Desulfurization Program carried out at the Jet Propulsion Laboratory, California Institute of Technology, and was sponsored by the U.S. Department of Energy through Interagency Agreement No. ET-77-I-01-9060 with NASA.

LITERATURE CITED

1.   Kalvinskas, J.J., et al., "Final report for phase I - coal desulfurization by low temperature chlorinolysis," Jet Propulsion Laboratory Publication 78-8, November 23, 1977.

2.   Kalvinskas, J.J., et al., "Final report coal desulfurization by low temperature chlorinolysis phase II," JPL Publication 80-15, January 15, 1980.

Table 1.   Selected Raw Coals Analyses

ERDA PSOC No.	Seam, County, State (Mine, Mesh Size)	Rank	Run No.	Sulfur Forms (Wt. %)[a]				Proximate Analysis (Wt. %)[a]					Ultimate Analysis (Wt. %)[a,c]						
				Organic	Pyritic	Sulfate	Total	Volatile Matter	Ash	Fixed Carbon	Heating Value (BTU/lb)[a]	Moisture[b]	C	H	S	N	Cl	O (by diff.)	Ash
276[d]	Ohio No. 8, Harrison, Ohio (Georgetown #24 Mine, 100-200 mesh)	Bit, HVA	1,5,6-7, 9-14,16																
	Average			1.17	2.65	0.06	3.87	37.2	11.5	51.3	12,755	1.89	71.6	5.67	3.91	1.28	0.16	5.87	11.5
	Std.Dev.			±0.16	±0.22	±0.06	±0.08	±0.5	±0.8	±0.5	±271	±0.37	±0.4	±0.41	±0.05	±0.19	±0.08	±1.19	±0.8
282[d]	Ill. No. 6, Jefferson, Ill. (Orient No. 6 Mine Washed[e], 100-200 mesh)	Bit, HVB	15,17, 20-25, 35,36, 38																
	Average			0.72	0.77	0.10	1.59	35.7	6.79	57.5	13,322	4.73	75.3	5.35	1.61	1.71	0.50	8.67	6.79
	Std.Dev.			±0.06	±0.03	±0.03	±0.07	±1.1	±0.42	±0.7	±37	±0.66	±0.9	±0.32	±0.04	±0.03	±0.04	±0.98	±0.42
282[d]	Ill. No. 6, Jefferson, Ill. (Orient No. 6 Mine, Washed[e], -1/8 in. to 100 mesh)		26,27, 33,34																
	Average			0.71	0.79	0.06	1.55										0.58		
	Std.Dev.			±0.04	±0.07	±0.03	±0.09										±0.11		
219	KY. No. 4, Hopkins, KY. (100-200 mesh)	Bit, HVB	28,29																
	Average			0.81	0.73	0.60	2.15	35.5	6.42	58.1	12,966	3.74	73.8	5.52	2.16	1.76	0.12	10.46	6.42
	Std.Dev.			±0.06	±0.01	±0.04	±0.01										±0.02		
026	Ill. No. 6, Saline, Ill. (100-200 mesh)	Bit, HVC	30-32																
	Average			1.62	1.17	0.66	3.45	35.0	9.87	55.1	12,540	3.36	66.8	5.03	3.40	1.45	0.19	15.23	9.87
	Std.Dev.			±0.12	±0.12	±0.05	±0.07										±0.02		
Island Creek Coal Co., No. 9, Union, County, KY. (100-200 mesh)		Bit	40,43																
	Average			1.85	1.78	0.10	3.72	40.0	10.5	49.5	13,135	1.40	72.8	5.25	3.73	1.40	0.19	6.05	10.5
	Std.Dev.			±0.04	±0.01	±0.01	±0.02										±0.01		

a: Dry basis   c: H and O have been corrected for moisture   e: Unwashed coal is 2.2 wt. % total sulfure, 22 wt. % ash
b: As determined basis   d: Coal obtained directly from the mine   f: Standard deviation = $\sqrt{\Sigma(x_i - \bar{x})^2/(N - 1)}$

Table 2.   Size Distribution of Ground and Sieved Raw Coals

Coal PSOC276
(Ground and Sieved to -100 +200 mesh)

Size Screening

Screen Mesh Size	Coal, Wt. %
retained on 60	0.3
60-80	0.2
80-100	0.2
100-120	10.2
120-170	54.7
170-200	17.7
200-270	10.2
270-325	1.0
passes 325	5.7

Coal PSOC 282
(Ground and Sieved to -1/8 inch to +100 mesh)

Screen Mesh Size	Coal, Wt. %
retained on 16	7.0
16-35	25.5
35-60	22.3
60-80	17.2
80-100	7.8
100-200	14.8
200-325	3.6
passes 325	1.8

Table 3.   Coal Desulfurization Data-Batch
Reactor

Table 4.   Change in Proximate Analyses of
Coals by Desulfurization

Composition (Wt. %) Coal PSOC No.	Raw Coal (Wt. %)	Desulfurized & Dechlorinated Coal (Wt. %)	Change	
			(Wt. %)	(%)
Volatile Matter				
276	37.2 ± 0.5	25.0	-12.2	-32.7
282	35.7 ± 1.1	26.0	-9.6	-27.0
219	35.5	27.7	-7.8	-22.0
026	35.0	26.2	-8.8	-25.2
Island Creek	40.0	24.8	-15.2	-38.0
Average	36.7	25.9	-10.7	-29.0
Ash				
276	11.5 ± 0.8	9.09	-2.41	-20.9
282	6.79 ± 0.4	6.47	-0.32	-4.7
219	6.42	4.70	-1.72	-26.8
026	9.87	7.58	-2.29	-23.2
Island Creek	10.5	5.81	-4.69	-44.7
Average	9.02	6.73	-2.29	-24.1
Fixed Carbon				
276	51.3 ± 0.5	65.8	+14.5	+28.4
282	57.5 ± 0.7	67.8	+10.3	+17.9
219	58.1	67.6	+9.5	+16.3
026	55.1	66.2	+11.1	+20.1
Island Creek	49.5	69.4	+19.9	+40.2
Average	54.3	67.4	+13.0	+24.6
Heating Value (B.T.U./lb)				
276	12,755 ± 271	12,649	-105	-0.8
282	13,322 ± 37	13,334	-987	-7.4
219	12,966	12,309	-657	-5.1
026	12,540	11,907	-633	-5.0
Island Creek	13,135	12,072	-1063	-8.1
Average	12,949	12,254	-689	-5.3

Reference: Coal PSOC 276 - Runs 3,9,14; PSOC 282 - Runs 18,19,20,
21,22; PSOC 219 - Run 29; PSOC 026 - Run 30, Island
Creek Coal - Run 42.

Table 5.   Change in Ultimate Analyses of
Coals by Desulfurization Process

Composition (Wt. %)/ Coal PSOC No.	Raw Coal (Wt. %)	Desulfurized & Dechlorinated Coal (Wt. %)	Change	
			(Wt. %)	(%)
Carbon				
276	71.6 ± 0.4	74.3	+2.66	+3.7
282	75.3 ± 0.9	75.0	-0.3	-0.4
219	73.8	75.7	+1.9	+2.6
026	66.8	73.3	+6.5	+9.7
Island Creek	72.8	74.8	+2.0	+2.7
Average	72.1	74.6	+2.5	+3.7
Hydrogen				
276	5.67 ± 0.41	4.19	-1.48	-26.1
282	5.35 ± 0.32	3.73	-1.62	-30.3
219	5.32	3.56	-1.76	-33.1
026	5.03	1.43	-3.60	-71.6
Island Creek	5.25	4.11	-1.14	-21.7
Average	5.32	3.40	-1.92	-36.6
Nitrogen				
276	1.28 ± 0.19	1.42	+0.14	+11.2
282	1.71 ± 0.03	1.53	-0.18	-10.5
219	1.76	1.56	-0.20	-11.4
026	1.45	1.33	-0.12	-8.3
Island Creek	1.40	1.38	-0.02	-1.4
Average	1.52	1.44	-0.08	-4.1
Oxygen by Difference				
276	5.87 ± 1.19	7.98	+2.1	+36.0
282	8.67 ± 0.98	10.58	+1.9	+22.0
219	10.46	11.4	+0.9	+9.0
026	13.33	9.29	-4.04	-30.3
Island Creek	6.05	9.92	+3.87	+64.0
Average	8.88	9.83	+0.95	+20.1
Chlorine				
276	0.16 ± 0.08	0.88	+0.72	+453
282	0.50 ± 0.04	2.17	+1.67	+335
219	0.12 ± 0.02	2.36	+2.24	+1867
026	0.19 ± 0.02	3.34	+3.15	+1658
Island Creek	0.19 ± 0.01	2.13	+1.94	+1021
Average	0.232	2.18	+1.94	+1067
Sulfur				
276	3.87 ± 0.08	1.93	-1.94	-50.2
282	1.59 ± 0.07	0.81	-0.78	-49.0
219	2.15 ± 0.01	0.74	-1.41	-65.6
026	3.45 ± 0.07	1.43	-2.02	-58.6
Island Creek	3.72 ± 0.02	1.74	-1.98	-53.2
Average	2.96	1.33	-1.63	-55.3

Reference: PSOC 276 - Runs 3,9,14; PSOC 282 - Runs 18,19,20,21,22;
PSOC 219 - Run 29; PSOC 026 - Run 30; Island Creek
Coal - Run 42.

Table 6.   Average Change in Ash Elemental
Composition for Five Bituminous
Coals by Desulfurization Process
(Coals PSOC No. 5, 278, 282, 219,
026, Island Creek)

Ash Element[a]	Raw Coal Ash Element (Wt. %)	Processed Coal Ash Element (Wt. %)	Ash Element Removal	
			(Wt. %)	(%)
Si	25	29	-4	-16
Fe	22	9	13	59
Al	13	14	0	0
Ca	1.7	0.5	1.2	71
K	2.1	2.1	0	0
S	1.1	0.5	0.6	55
Ti	0.7	0.8	-0.1	-14
Na	0.7	0.5	0.2	29
Mg	0.5	0.5	0	0
P	0.04	0.06	0	0
Mn	0.02	0.02	0	0
Ash	9.0	7.1	1.8	20

a - Element concentrations regarded as wt. % of ash are the average of the five
bituminous coals analyzed.

Table 7. Change in Trace Element Composition
of Coal by Desulfurization Process

Element(ppm)/ Coal PSOC No.	Raw Coal (ppm)	Processed Coal (ppm)	Removal (ppm)	(%)
Lead				
276	111	155	-44	-40
282	197	127	70	+35
026	<50	<50	-	-
Island Creek	<50	<50	-	-
Average	77	95	18	23
Arsenic				
276	6	3	3	+50
282	9	3	6	+67
026	3	1	2	+67
Island Creek	3	4	-1	-33
Average	5	3	2	40
Selenium				
276	5	6	-1	-20
282	4	5	-1	-25
026	8	6	2	25
Island Creek	3	4	-1	-33
Average	5	5	-1	-20
Mercury				
276	0.2	0.2	0	0
282	0.6	0.1	0.5	83
026	0.2	<0.1	0.1	50
Island Creek	0.4	0.02	0.38	95
Average	0.3	0.1	0.2	67
Vanadium[a]				

[a]Present at less than detection limit of 50 ppm.

Table 8. Batch Reactor Feed Composition

Run Nos.	Coal PSOC No.	Raw Coal	Organic Fraction	Ash	Cl⁻	Sulfur	Chlorine	Methyl-chloroform
			Raw Coal Composition					
1,3,7,10	276	2000	1694	226	3.2	77.2	680-1320	4000
4,5,6,8,9,12,13,14,16	276	2000	1694	226	3.2	77.2	390-990	0
17,19,21,23,25,38,39	282	2000	1820	135	10.1	31.7	880-1660	4000
15,18,22,24,35,36,37	282	2000	1820	135	10.1	31.7	650-1660	0
27,34	282C[a]	2000	1792	157	11.6	31.0	1050-1370	4000
26,33	282C[a]	2000	1792	157	11.6	31.0	1330-1560	0
28	219	1656	1511	107	2.4	35.6	1400	4000
29	219	1900	1734	123	2.4	40.9	1430	0
31	026	1706	1474	170	2.4	58.9	1380	4000
30,32	026	1698	1467	169	2.4	58.6	1260-1620	0
43,44	Island Creek	1566-2000	1202-1716	99-206	4-238	26-74	1380	4000
40,41,42	Island Creek	1692-2000	1297-1716	89-206	4-278	28-45	680-790	0

[a]PSOC 282C particle is 16 x 100 mesh. Nominal size is 100 x 200 mesh.

Table 9. Coal Filtrate and Wash Water Analyses

Run Nos.	Coal PSOC No.	Product Coal[a] (grams)	Filtrate & Wash-water (liters)	Cl⁻ g/ℓ	Cl⁻ Coal Wt.%	SO₄⁼ g/ℓ	SO₄⁼ Coal Wt.%	Fe³⁺⁺ g/ℓ	Fe³⁺⁺ Coal Wt.%	Ca⁺⁺ g/ℓ	Ca⁺⁺ Coal Wt.%	Na⁺ g/ℓ	Na⁺ Coal Wt.%	Aℓ⁺⁺ g/ℓ	Aℓ⁺⁺ Coal Wt.%	Mg⁺⁺ g/ℓ	Mg⁺⁺ Coal Wt.%	K⁺ g/ℓ	K⁺ Coal Wt.%
1,3,7,10[c]	276	1489	13.3	39.7	35.5	8.2	7.3	7.0	6.3	0.25	0.22	0.14	0.12	0.04	0.04	0.11	0.10	0.01	0.009
4,5,6,8,9 12,13,14,16[d]	276	1672	8.6	66.5	34.2	8.8	4.5	8.1	4.2	0.43	0.22	0.14	0.07	0.10	0.05	0.07	0.04	0.02	0.01
17,19,21,23, 25,38,39c	282	1473	13.5	34.1	31.2	2.0	1.9	2.1	1.9	0.09	0.08	0.15	0.14	0.04	0.04	0.02	0.02	0.01	0.01
15,18,22,24, 35,36,37d	282	1824	10.9	78.6	47.0	2.9	1.7	3.9	2.3	0.13	0.08	0.20	0.12	0.07	0.04	0.03	0.02	0.01	0.01
27,34[c]	282C	1394	10.0	35.0	25.1	1.8	1.3	3.4	2.4	0.11	0.08	0.11	0.08	0.05	0.04	0.04	0.03	0.01	0.01
26,33[d]	282C	1504	12.2	42.0	34.1	1.7	1.4	2.4	1.9	0.11	0.09	0.12	0.10	0.03	0.03	0.03	0.02	0.01	0.01
28[c]	219	1344	9.8	49.2	35.9	4.5	3.3	4.6	3.3	0.05	0.04	0.03	0.02	0.09	0.07	0.04	0.03	0.02	0.01
29[d]	219	1409	4.8	124.1	42.3	8.7	3.0	10.9	3.7	0.09	0.03	0.06	0.02	0.93	0.32	0.07	0.02	0.03	0.01
31[c]	026	1020	9.1	42.4	37.8	6.0	5.3	4.9	4.4	0.07	0.06	0.04	0.04	0.11	0.10	0.03	0.03	0.03	0.03
30,32[d]	026	1513	6.3	116.2	48.4	11.1	4.6	12.3	5.1	0.09	0.04	0.04	0.02	0.41	0.17	0.05	0.02	0.04	0.02
43,44[c]	Island Creek	1418	21.7	28.2	43.1	2.0	3.1	3.4	5.2	0.14	0.21	0.04	0.07	0.03	0.05	0.02	0.03	0.01	0.01
40,41,42[d]	Island Creek	1795	19.7	23.1	25.3	1.5	1.3	1.8	2.0	0.17	0.19	0.06	0.06	0.05	0.04	0.03	0.03	0.01	0.01

a: Processed coal, dry weight (not including coal in equipment water rinse).
b: Coal weight percent based on dry weight of processed coal cited in Note a.
c: Runs with methylchloroform in coal slurry.
d: Runs with water and no organic solvent in coal slurry.

Table 10. Material Balance - Chlorination, Solvent Recovery and Coal Filtration - Wash (Coal, Solvent, Chlorine and Sulfur Accounting)

Run Nos.	Raw Coal grams	Raw Coal %	Organic Fraction grams	Organic Fraction %	Ash grams	Ash %	Sulfur grams	Sulfur %	Chlorine grams	Chlorine %	Methyl-Chloroform grams	Methyl-Chloroform %
1,3,7,10	2024	101.2	1709	100.5	246	109	72.4	94	613-847	77	3290	82
4,5,6,8,9,12,13,14,16	2024	101.2	1708	100.4	245	108	67	86	402-1390	98	0	--
17,19,21,23,25,38,39	1917	95.8	1515	94.8	145	107	14	87	398-1452	83	3320	83
15,18,22,24,35,36,37	2040	102	1775	101	150	111	18	92	397-1161	74	0	--
27,34	2025	101	1800	100	185	118	30	98	1083-1108	92	3200	80
26,33	2025	101	1800	100	185	118	30	96	805-1529	80	0	--
31	1609	94	1381	94	175	103	51	87	1232	89	3041	76
30,32	1728	100	1462	100	209	124	56	95	1162-1425	90	0	--
43,44	1604-2154	105	1164-1847	105	119-179	147	25-68	93	1106-1378	83	3800	95
40,41,42	1800-2019	101	1307-1733	100	124-211	131	35-71	102	800-951	99	0	--
Average	2000	100 ±3	1928	99 ±3		114		94		86	3320	83 ±7

Table 11. Coal Dechlorination Data

Run Nos.	Coal PSOC No.	Dechlorination Time (min)	Average Input Coal Mass[a] (grams)	Average Input Coal Chlorine Content Ave. (wt. %)	Average Input Coal Chlorine Content Range (wt. %)	Average Recovered Coal Mass[a] (grams)	Average Recovered Coal Chlorine Content Ave. (wt. %)	Average Recovered Coal Chlorine Content Range (wt. %)	Chlorine Removed (%)	Product Coal Recovery (%)
1-14, 16	276	60,90	1323	6.26	1.53-10.8	1119	0.98	0.23-1.48	87.0	89.3
15,17-25, 35,37,39	282 (100 x 200 Mesh)	90	1372	14.87	3.2-34.4	1032	1.91	0.44-3.84	90.3	86.7
26,27,33,34	282 (-1/8 x 100 Mesh)	90	1347	15.22	13.2-17.5	1054	1.63	0.86-2.23	91.6	91.1
28,29	219	60-65	1109	18.5	17.5-19.6	923	1.56	1.44-1.68	93.0	100.
30,32	026	60-90	1292	16.8	15.8-17.9	902	1.68	1.23-2.13	93.0	82.5
42,44	Island Creek	90-95	1078	25.0	16.7-33.2	756	3.74	2.37-5.12	89.5	90.0

Condition: Dechlorination temperature 400°C, tube rotation 4 RPM, preheat time 20 minutes, nitrogen purge rate of 30 SCFH, coal mesh size 100 x 200)

a - dry basis

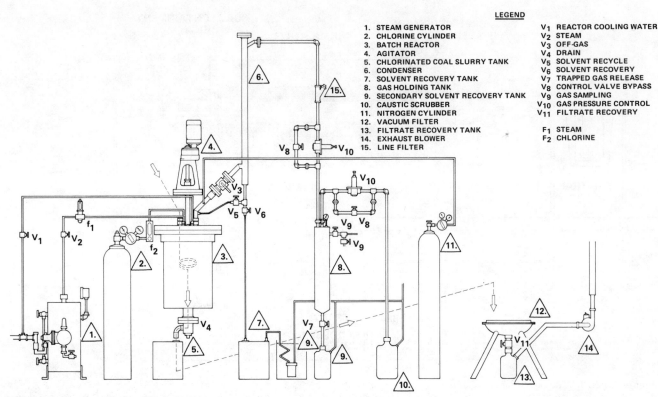

LEGEND

1. STEAM GENERATOR
2. CHLORINE CYLINDER
3. BATCH REACTOR
4. AGITATOR
5. CHLORINATED COAL SLURRY TANK
6. CONDENSER
7. SOLVENT RECOVERY TANK
8. GAS HOLDING TANK
9. SECONDARY SOLVENT RECOVERY TANK
10. CAUSTIC SCRUBBER
11. NITROGEN CYLINDER
12. VACUUM FILTER
13. FILTRATE RECOVERY TANK
14. EXHAUST BLOWER
15. LINE FILTER

V_1 REACTOR COOLING WATER
V_2 STEAM
V_3 OFF-GAS
V_4 DRAIN
V_5 SOLVENT RECYCLE
V_6 SOLVENT RECOVERY
V_7 TRAPPED GAS RELEASE
V_8 CONTROL VALVE BYPASS
V_9 GAS SAMPLING
V_{10} GAS PRESSURE CONTROL
V_{11} FILTRATE RECOVERY

F_1 STEAM
F_2 CHLORINE

Figure 1. Bench-scale batch reactor system.

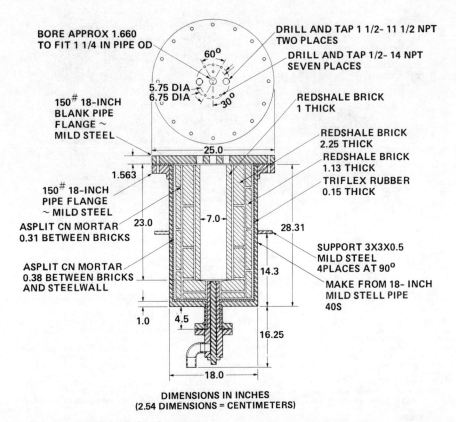

Figure 2. Bench-scale batch reactor.

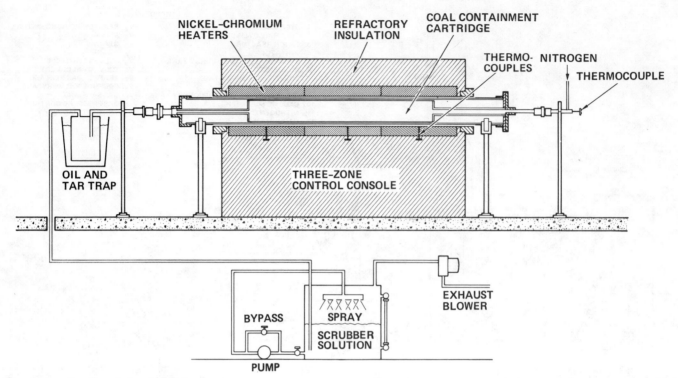

Figure 3. Bench-scale batch dechlorinator.

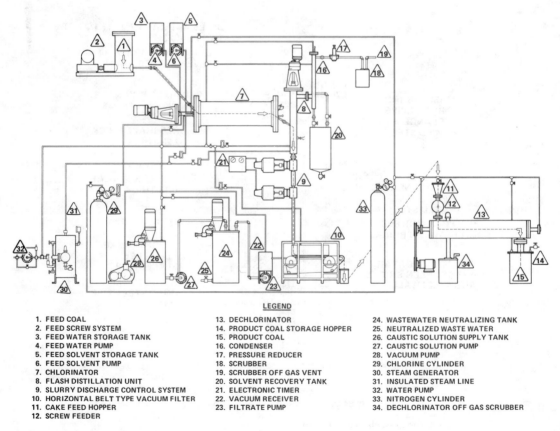

LEGEND

1. FEED COAL	13. DECHLORINATOR	24. WASTEWATER NEUTRALIZING TANK
2. FEED SCREW SYSTEM	14. PRODUCT COAL STORAGE HOPPER	25. NEUTRALIZED WASTE WATER
3. FEED WATER STORAGE TANK	15. PRODUCT COAL	26. CAUSTIC SOLUTION SUPPLY TANK
4. FEED WATER PUMP	16. CONDENSER	27. CAUSTIC SOLUTION PUMP
5. FEED SOLVENT STORAGE TANK	17. PRESSURE REDUCER	28. VACUUM PUMP
6. FEED SOLVENT PUMP	18. SCRUBBER	29. CHLORINE CYLINDER
7. CHLORINATOR	19. SCRUBBER OFF GAS VENT	30. STEAM GENERATOR
8. FLASH DISTILLATION UNIT	20. SOLVENT RECOVERY TANK	31. INSULATED STEAM LINE
9. SLURRY DISCHARGE CONTROL SYSTEM	21. ELECTRONIC TIMER	32. WATER PUMP
10. HORIZONTAL BELT TYPE VACUUM FILTER	22. VACUUM RECEIVER	33. NITROGEN CYLINDER
11. CAKE FEED HOPPER	23. FILTRATE PUMP	34. DECHLORINATOR OFF GAS SCRUBBER
12. SCREW FEEDER		

Figure 4. Equipment flow Schematic—mini-pilot-plant.

THE EFFECT OF FUEL-SULFUR ON NO$_x$ FORMATION IN A REFRACTORY BURNER

S.-K. TANG
STUART W. CHURCHILL
and
NOAM LIOR

University of Pennsylvania
Philadelphia, Pennsylvania 19104

Combustion inside a refractory tube is stabilized by thermal conduction in the tube wall and by wall-to-wall radiation. In the turbulent regime, plug flow is approached closely and backmixing is negligible. Hence, the process is ideal for kinetic studies. In this investigation the effect of adding up to 5.47% wt. tertiary butyl mercaptan and up to 5.17% wt. pyridine to hexane has been investigated experimentally and theoretically. The mercaptan was found to reduce the thermal NO$_x$ slightly and the fuel-NO$_x$ significantly for the combustion of pre-evaporated droplets. The theoretical results are in qualitative agreement with the experimental measurements and provide considerable insight regarding the paths of reaction.

The work is part of a continuing investigation by Churchill and coworkers of combustion inside a refractory tube. In this process of combustion the thermal feedback for the stabilization of the flame is by conduction in the tube wall and by wall-to-wall radiation. In the turbulent regime, backmixing is negligible and the radial profiles of velocity, temperature and composition are nearly flat. Thus the behavior closely approaches the idealized case of plug flow, greatly simplifying both experimental modelling and the analyses of experimental data.

The flame front is essentially a step in temperature and composition corresponding to conversion of the fuel to CO and H$_2$O. The temperature then rises more slowly to a maximum as the CO is oxidized to CO$_2$. The exit temperature of the burned gas is very close to the theoretical equilibrium value owing to minimal heat losses from the combustion chamber, primarily by radiation out the end of the tube. The peak temperature of the gas stream is slightly above the exit value, owing to the internal thermal feedback. The flow in the post-flame zone is nearly adiabatic, and after the peak temperature is nearly isothermal as well.

S.-K. Tang is now with Shell Development Co., Houston, TX.

0065-8812-81-4720-0211-$2.00.
© The American Institute of Chemical Engineers, 1981

Owing to the great thermal inertia of the ceramic wall, the flame is very stable and free from oscillations. Also, owing to the combined radiative and conductive feedback, multiple stationary states have been found to exist for the same external conditions.

Chen and Churchill (1) determined experimentally the range of conditions under which flames of premixed propane and air can be stabilized inside 4.76 and 9.52 mm tubes. They (2) also developed a theoretical model using global kinetics which agreed with their experimentally observed limits of stability and their measured wall temperature profiles. The model predicted the existence of additional stationary states. Bernstein and Churchill (3) confirmed the existence of the predicted multiple stationary states and observed that the formation of thermal NO$_x$ in this burner was exceptionally low (5 to 32 ppm) and linear with residence time in the nearly isothermal post-flame zone.

Choi and Churchill (4) obtained similar exit-gas compositions for uniformly sized droplets of hexane which were completely evaporated by the thermal feedback prior to ignition. They (5) improved the model and method of solution of Chen and Churchill (2) and extended it to include atomized droplets of liquid fuel. Goepp et al.(6) confirmed the existence of multiple stationary states with hexane droplets and showed that these states could be used to vary the NO$_x$ and CO contents of the exit-gas for the same inlet conditions. Tang et al. (7) utilized the thermal inertia of the system to study the combustion of hexane droplets at pseudostationary states, and hence to vary the post-flame residence time for fixed external conditions. They found NO_x from fuel-nitrogen to be independent of the post-flame residence time, and therefore presumably to be formed in or near the flame front. Tang and Churchill (8,9) used detailed chemical kinetic models to predict the composition of the burned gas for both hexane, and hexane doped with nitrogen compounds.

In the present work a sulfur compound as well as a nitrogen compound has been added to the fuel. Rate mechanisms involving sulfur have correspondingly been added to the previous model. Prior work on the interaction of nitrogen and sulfur has been rather limited. Wendt and Ekmann (10) showed that high levels of SO_2 or H_2S in premixed flat flames inhibited the formation of thermal NO_x. Wendt et al. (11), using the same process, observed both enhancement and inhibition of the formation of fuel-NO_x, depending on the equivalence ratio and the residence time. de Soete (12) obtained similar results for prompt NO_x and fuel-NO_x. These three studies were limited to fuel-rich mixtures.

EXPERIMENTAL APPARATUS AND PROCEDURE

The apparatus, operational procedure, and methods of chemical analysis used in this work were essentially the same as that used by Goepp et al. (6) and Tang et al. (7) and hence will only be described functionally. The evaporation and combustion chamber consisted of a round straight channel 9.7 mm in diameter and 1221 mm long, of which the first 970 mm were a thermally insulated aluminum oxide tube and the final 251 mm the central hole in a cast alumina block. This central hole was surrounded by six outer holes in which a propane-air mixture was burned to minimize radial heat losses.

A single chain of uniformly sized and uniformly spaced droplets was generated by passing a metered stream of reagent-grade hexane through a hypodermic needle vibrated at 5kHz. Compressed air was filtered, metered, preheated and introduced into the upper end of the evaporation and combustion chamber, just above the hexane inlet.

Small concentrations of pyridine and tertiary butyl mercaptan were mixed separately and jointly with the hexane to simulate nitrogen- and sulfur-containing fuels.

The temperature profile along the inner wall of the evaporation tube and combustion channel was measured with a series of thermocouples near the surface. A sample of the gas stream leaving the combustion chamber was pulled through a fused-quartz tube, cooled, subcooled for water removal, and analyzed for CO, CO_2, O_2, NO and NO_x. In view of the findings of Johnson et al. (13) and others, the measured NO_{2+} is presumed to be formed in the sampling tube. Hence only values of NO_x are reported, and they are presumed to represent NO. Unburned fuel has been asserted to influence the determination of NO_x by thermal conversion to NO. The values of NO_x for fuel-rich mixtures may be in some error for this reason. However, the agreement between the experimental and theoretical result in Figure 11 of reference (8) for thermal NO_x up to $\phi = 1.1$, and in Figure 1 of reference (9) for fuel-NO_x up to $\phi = 1.3$, suggests that this error may not be appreciable, at least for those conditions.

Owing to the thermal inertia of the ceramic combustion chamber, many hours are required to establish a true equilibrium state. However, this thermal inertia made possible the establishment of pseudo-stationary flames at the same fixed location for different rates of flow of air and fuel, whereas a separate location for each would be attained at equilibrium. To vary the equivalence ratio for fuel-lean mixtures the rate of flow of air was varied for a nearly fixed rate of flow of fuel, while for the fuel-rich mixtures the rate of flow of fuel was varied for a fixed rate of flow of air. When the rate of flow of fuel was varied, the diameter and initial spacing of the droplets varied. The diameters ranged from 215 to 291 µm, and the spacings from 542 to 1261 µm. Since in all cases, the droplets evaporated completely prior to ignition, the diameter and spacing are presumed to have had no effect on the results other than to change the equilibrium location of the flame front.

The experimental flame-front location was arbitrarily defined as the point of inflection in a plot of the wall temperature versus distance along the combustion channel. The gas temperature profile was not measured but was computed from a theoretical model, as described below, using the measured wall-temperature profile as an input.

All of the runs reported herein were for a flame-front location approximately 220 mm from the exit. The total rate of flow varied with the equivalence ratio, as indicated above, but the results with and without fuel-nitrogen and fuel-sulfur are directly comparable for any particular equivalence ratio. Furthermore, because of the compensating effects of temperature and flow rate, the residence times in the post-flame zone were all confined to the narrow range of 6.8 ± 0.8 ms for all equivalence ratios.

EXPERIMENTAL RESULTS

The addition of 5.47% wt. tertiary butyl mercaptan (or 1.94% wt. sulfur) is seen in Figure 1 to produce only a slight decrease in the formation of thermal NO_x. The data for pure hexane are from Tang et al. ([7]).

The same amount of mercaptan is seen in Figure 2 to reduce significantly the formation of NO_x from hexane which also contains 5.17% wt. pyridine (or 0.92% wt. nitrogen). The dotted curve in Figure 2 represents total conversion of the fuel-nitrogen to NO_x plus the

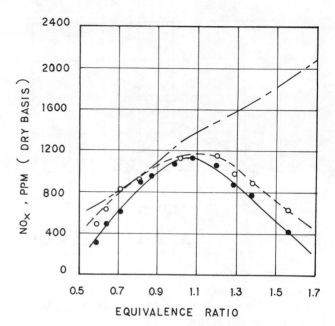

Figure 2.   Measured NO_x from Combustion of Hexane Droplets Doped with Nitrogen and Sulfur.

Post-flame residence time = 6.8 ± 0.8 ms

--o--   hexane with 5.17% wt. pyridine
--●--   hexane with 5.17% wt. pyridine and 5.17% wt. tertiary butyl mercaptan
----   complete conversion of fuel-nitrogen plus estimated thermal NO_x

thermal NO_x from Figure 1. Similar behavior for 2.2% wt. pyridine (or 0.39% wt. nitrogen) and 0.57% wt. pyridine (or 0.10% wt. nitrogen) is apparent in Figures 3 and 4. The reduction of the ppm of NO_x is about the same in Figures 2 and 3 but is considerably less in Figure 4.

The effect of adding 2.1% wt. of the mercaptan (or 0.75% wt. sulfur) to hexane with 2.2% wt. and 0.57% wt. pyridine is indicated in Figures 5 and 6. Comparison of Figures 3 and 5 for 2.2% wt. pyridine indicates that 2.1% wt. of the mercaptan is just as effective as 5.47% wt. Comparison of Figures 4 and 6 for 0.57% pyridine indicates a greater effectiveness for 5.47% wt. of the mercaptan for an equivalence ratio near unity but reductions similar to those of Figures 3 and 5 for very lean and very rich mixtures.

The CO content of the burned gas did not appear to be influenced significantly by the addition of fuel-sulfur and/or fuel-nitrogen.

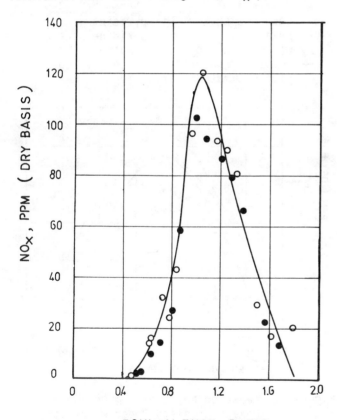

EQUIVALENCE RATIO

Figure 1.   Effect of Fuel-Sulfur on Thermal NO_x from Combustion of Hexane Droplets
Post-flame residence time = 6.8 ± 0.8 ms

o - pure hexane
● - hexane with 5.47% wt. tertiary butyl mercaptan

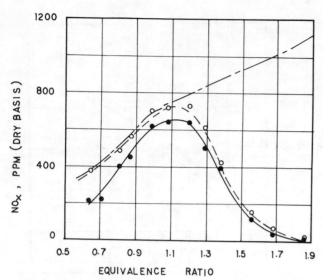

Figure 3.   Measured NO_x from Combustion of Hexane Droplets Doped with Nitrogen and Sulfur.

Post-flame residence time = 6.8 ± 0.8 ms

--o--   hexane with 2.2% wt. pyridine

—●—   hexane with 2.2% wt. pyridine and 5.47% wt. tertiary butyl mercaptan

---   complete conversion of fuel-nitrogen plus estimated thermal NO_x

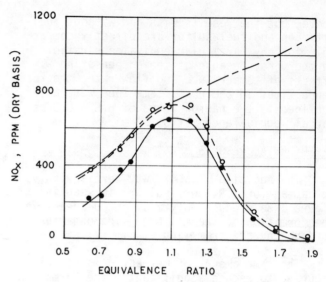

Figure 5.   Measured NO_x from Combustion of Hexane Droplets Doped with Nitrogen and Sulfur.

Post-flame residence time = 6.8 ± 0.8 ms

--o--   hexane with 2.2% wt. pyridine

—●—   hexane with 2.2% wt. pyridine plus 2.1% wt. tertiary butyl mercaptan

---   complete conversion of fuel-nitrogen plus estimated thermal NO_x

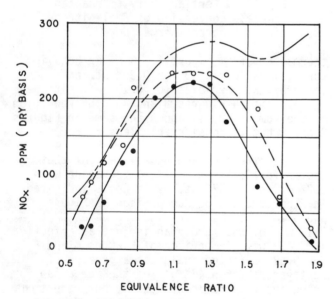

Figure 4.   Measured NO_x from Combustion of Hexane Droplets Doped with Nitrogen and Sulfur.

Post-flame residence time = 6.8 ± 0.8 ms

--o--   hexane with 0.57% wt. pyridine

—●—   hexane with 0.57% wt. pyridine and 5.47% wt. tertiary butyl mercaptan

---   complete conversion of fuel-nitrogen plus estimated thermal NO_x

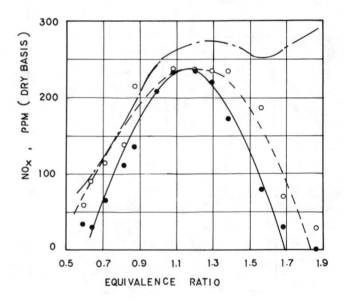

Figure 6.   Measured NO_x from Combustion of Hexane Droplets Doped with Nitrogen and Sulfur.

Post-flame residence time = 6.8 ± 0.8 ms

--o--   hexane with 0.57% wt. pyridine

—●—   hexane with 0.57% wt. pyridine plus 2.1% wt. tertiary butyl mercaptan

---   complete conversion of fuel-nitrogen plus estimated thermal NO_x

THEORETICAL MODEL

The theoretical model of Tang and Churchill (8,9) was extended for fuel-sulfur by adding the 16 reversible reaction mechanisms of Table I to the 44 used for the sulfur-free case. The rate constants for the forward reactions are included in Table I. The equilibrium constants for these reactions, as a function of temperature, were taken from the JANAF tables (14). A documentation of the sources of these rate expressions and constants is given by Tang (15). The choice of these particular mechanisms and rate constants cannot readily be justified. They are merely asserted to be representative. Improvement depends on kinetic rather than on combustion studies.

For the preflame zone, the prior model for the combustion of pure hexane consisted of Eulerian differential mass, momentum and energy balances for a single droplet, and Eulerian differential mass, energy and hexane balances for the gas stream. A single global reaction mechanism was utilized for the oxidation of the hexane to CO and H_2O. The flame front was defined as the point where the mole fraction of hexane fell below 1×10^{-6}. In the post flame zone, an Eulerian differential balance was written for each species; adiabatic flow was postulated for simplicity. The model was integrated stepwise by finite differences. For the calculations herein the hydrogen associated with the fuel-sulfur as H_2S and the hydrogen and carbon associated with the fuel-nitrogen as HCN were deleted from the fuel oxidized to CO and H_2O in the preflame zone. These amounts of H_2S and HCN were then postulated to be formed instantaneously at the flame front.

For fuel-rich mixtures hydrocarbon fragments are known to be formed in the flame and to persist in the post-flame region. Hence,

TABLE I

Added Mechanisms and Forward Rate Constants for Sufur

$$A + B \rightleftarrows C + D$$

No.	A	B	C	D	k $m^3/mol \cdot s$	E/R K
1	SO_3	O	SO_2	O_2	6.5 E8	5438
2	SO_2+M	O	SO_3	M	1.0 E11*	0
3	SO_2	NO_2	NO_2	SO_3	1.0 E7	13600
4	SO	O_2	SO_2	O	4.0 E9	0
5	SO	OH	SO_2	H	1.0 E8	0
6	SO	NO_2	SO_2	NO	1.0 E7	0
7	SO+M	O	SO_2	M	1.0 E12*	0
8	SO	N	NO	S	5.1 E5	0
9	SO	SO	SO_2	S	1.0 E5	0
10	H_2S	O	OH	SH	1.0 E5	0
11	H_2S	OH	H_2O	SH	1.0 E5	0
12	H_2S	H	H_2	SH	1.0 E9	13600
13	SH	O_2	OH	SO	1.0 E8	0
14	COS	O	CO	SO	1.0 E6	0
15	COS	O	CO_2	S	1.0 E8	2770
16	S	O_2	SO	O	1.0 E7	2820

*$m^6/mol \cdot s$

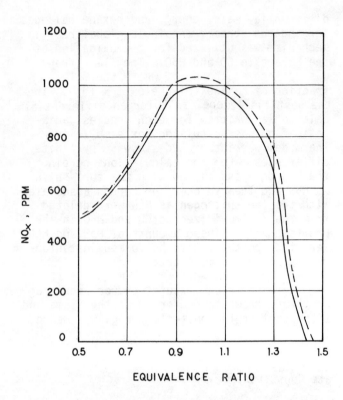

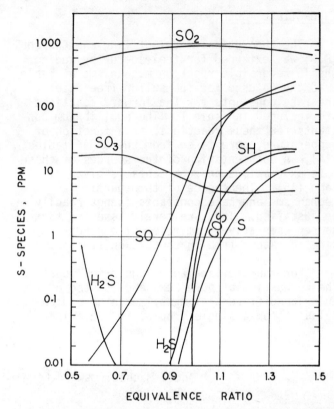

Figure 7.   Computed NO_x from Combustion of Hexane Droplets Doped with Nitrogen and Sulfur.

Post-flame residence time = 6.8 ± 0.8 ms

---- hexane with 4.15% pyridine
——— hexane with 4.15% wt. pyridine and 5.12% wt. tertiary butyl mercaptan

the global model used herein for the preflame and flame zones, including the complete conversion of nitrogen and sulfur to HCN and H_2S, might be expected to predict compositions in increasing error for all respects for equivalence ratios greater than 1.0, and particularly for ratios greater than 1.37, for which there is insufficient oxygen even to convert the hexane to CO and H_2O. However, the previously mentioned agreement between the experimental and theoretical values of NO_x in references (8) and (9) suggests that the error may not be significant for Φ < 1.1 and perhaps even for φ < 1.3.

Illustrative calculations were carried out for hexane with 5.12% wt. of tertiary butyl mercaptan and with or without 4.15% wt. pyridine.

THEORETICAL RESULTS

The computed effect of fuel-sulfur on the

Figure 8.   Computed Concentrations of Sulfur-Containing Species from Combustion of Hexane Droplets Doped with 5.12% wt. Tertiary Butyl Mercaptan and 4.15% wt. Pyridine

Post-flame residence time = 6.8 ± 0.8 ms

formation of NO from fuel-nitrogen is illustrated in Figure 7. The residence time is in the same range but the concentrations of fuel-nitrogen and fuel-sulfur (as noted above) are slightly less than in the experimental work of Figure 2. The qualitative effect of the fuel-sulfur on NO_x formation is the same but the computed reduction is less than in the experiments. (The computed concentrations of NO_2 were in all cases less than 1 ppm. Hence no distinction need be made between NO and NO_x.)

The computed distribution of fuel-sulfur in the exit-gas stream is plotted versus the equivalence ratio in Figure 8. The corresponding equilibrium concentrations are shown in Figure 9. The computed concentration of SO_3 is relatively constant and considerably above the equilibrium value for all equivalence ratios. The computed concentration of H_2S is higher than the equilibrium value since it started out higher at the flame front, but is of significant magnitude only

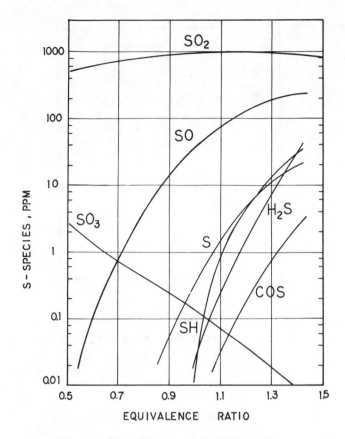

Figure 9.   Computed Equilibrium Concentrations
            of Sulfur-Containing Species for
            Combustion of Hexane Droplets Doped
            with 5.12% wt. Tertiary Butyl Mer-
            captan and 4.15% wt. Pyridine.

for fuel-rich mixtures. The SO concentration
approaches the equilibrium value and is of significant magnitude in the fuel-rich range.

The various nitrogen containing species,
other than NO and NO_2, are plotted in Figure 10.
HCN and NH_3 appear to become significant as the
equivalence ratio increases above unity. However, these high values of HCN may be merely
an artifact of the postulate of complete decomposition of the pyridine to HCN in the flame
front. The effect of fuel-sulfur on all of the
species in this figure is seen to be slight.
However fuel-sulfur is seen in Figure 11 to
have significant effect on the concentrations
of O and OH, particularly for equivalence
ratios less than unity. The reduction in these
species undoubtedly contributes to the reduction in NO formation.

The computed histories of H_2O, SO, SO_2
and SO_3 in the post-flame zone are plotted in
Figure 12 for an equivalence ratio of 0.71.
The concentrations of all other sulfur-containing species are less than 1 ppm.   The

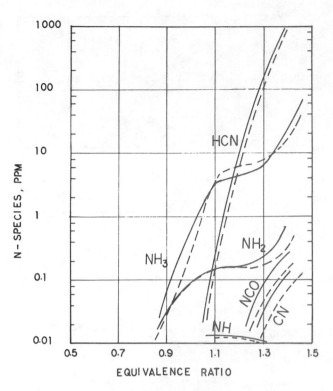

Figure 10.   Computed Concentrations of Nitro-
             gen-Containing Species (other than
             NO and NO_2) from Combustion of
             Hexane Droplets Doped with Nitro-
             gen and Sulfur.

Post-flame residence time = 6.8 ± 0.8 ms

---- hexane with 4.15% wt. pyridine
—— hexane with 4.15% wt. pyridine and 5.12%
     wt. tertiary butyl mercaptan

formation of SO_2 and SO_3 is seen to be essentially complete in less than a millisecond.
Figure 13 is a similar plot for an equivalence
ratio of 1.29. The formation of SO_2, SH, SO,
S and SO_3 are completed rapidly but the concentration of COS is still increasing significantly after 6 ms.

Figure 14 is a plot of the history of NO
in the post-flame zone for an equivalence
ratio of 1.29. In the absence of fuel-sulfur, the NO concentration rises rapidly to a
maximum value of 900 ppm in 0.3 ms then
slowly decreases to 790 ppm at 6 ms. Fuel-
sulfur reduces the peak value to about 860 ppm
and results in a concentration of 740 ppm at
6 ms. The effect of the fuel-sulfur on NO
thus appears to be confined to the period of
formation in the immediate post-flame zone.
For lean mixtures (not shown) NO rises rapidly
at first and then more slowly, the effect of
fuel-sulfur is almost negligible, and HCN decreases continuously with residence time.

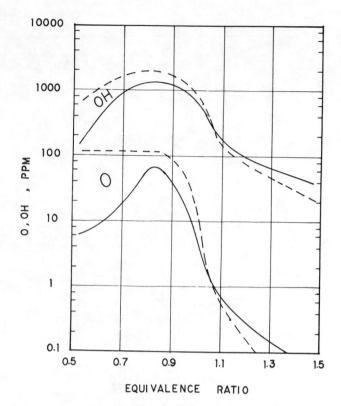

Figure 11.   Computed Concentrations of O and OH
             from Combustion of Hexane Droplets
             Doped with Nitrogen and Sulfur.

Post-flame residence time = 6.8 ± 0.8 ms

---- hexane with 4.15% wt. pyridine
——— hexane with 4.15% wt. pyridine and 5.12%
     wt. tertiary butyl mercaptan

CONCLUSIONS

The following conclusions can be drawn
from the experimental combustion in a refractory tube of hexane droplets doped with pyridine and/or tertiary butyl mercaptan. (The
post-flame residence time was approximately
6.8 ms in all of these tests).
   1) The addition of the mercaptan reduces
the formation of thermal NO_x only slightly
(see Figure 1).
   2) The addition of the mercaptan reduces
the formation of fuel-NO_x from pyridine
significantly.
   3) The reduction in fuel-NO_x was relatively independent of the fuel-sulfur concentration over a range from 0.75 to
1.94% wt.
   4) Fuel-sulfur and fuel-nitrogen do not
significantly influence the concentration
of CO in the burned gas.

The following additional conclusions follow from a theoretical model based on global

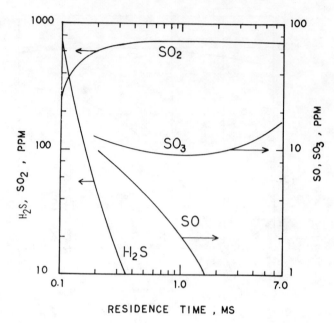

Figure 12.   Computed Post-flame History of H_2S,
             SO, SO_2 and SO_3 from Combustion of
             Hexane Droplets Doped with 4.15%
             wt. Pyridine and 5.12% wt. Ter-
             tiary Butyl Mercaptan

             Equivalence ratio = 0.71

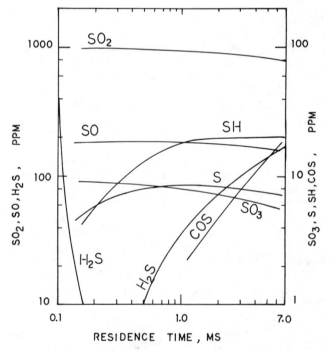

Figure 13.   Computed Post-flame History of
             Sulfur-Containing Species from
             Combustion of Hexane Droplets
             Doped with 4.15% wt. Pyridine and
             5.12% wt. Tertiary Butyl Mercaptan

             Equivalence ratio = 1.29

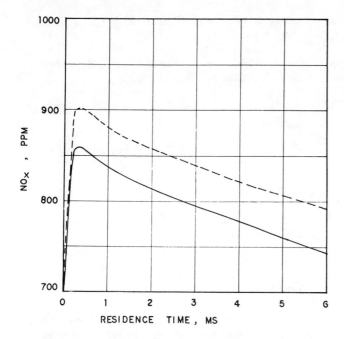

Figure 14.   Computed Post-flame Histories of NO from Combustion of Hexane Droplets Doped with Nitrogen and Sulfur.

Equivalence ratio = 1.29

---- hexane with 4.15% wt. pyridine
——— hexane with 4.15% wt. pyridine and 5.12% wt. tertiary butyl mercaptan

kinetics in the preflame zone but detailed reaction mechanisms for the post-flame zone. The computations were for the same conditions as the experiments except for the concentrations of pyridine and tertiary butyl mercaptan which were 4.15% wt. and 5.12% wt., respectively.

1) Prior studies (8,9) indicate that the predictions of thermal NO_x by this model are valid up to an equivalence ratio of 1.1 and of fuel-NO_x up to an equivalence ratio of 1.3.

2) The predicted effect of fuel sulfur on the formation of fuel-NO_x is in qualitative agreement with the measurement but the reductions are smaller. Some of the rate constants in the model are therefore presumed to be in error.

3) The computed concentrations of NO_2 are in all cases less than 1 ppm. Hence no distinction is necessary between the experimental values of NO_x and the computed values of NO. This result is in agreement with the experimental observations of Johnson et al. (10) and others.

4) The exit concentration of SO_3 is relatively independent of the equivalence ratio and far in excess of the equilibrium value.

5) The exit concentration of H_2S is higher than the equilibrium value (perhaps owing to the postulate of complete decomposition of the mercaptan of H_2S at the flame front) but is of significant magnitude only for fuel-rich mixtures.

6) The exit concentration of SO approaches the equilibrium value and is of significant magnitude for fuel-rich mixtures.

7) The exit concentrations of HCN and NH_3 become significant as the equivalence ratio increases above unity. However, these high values of HCN may be an artifact of the postulate of complete decomposition of pyridine to HCN in the flame front.

8) The effect of fuel-sulfur is negligible on the exit concentrations of nitrogen species other than NO.

9) Fuel-sulfur significantly reduces the exit concentrations of OH and O in the fuel-lean range. This reduction may be responsible for the reduction in NO.

10) The formation of SO_2 and SO_3 is essentially complete in less than a millisecond.

11) In fuel-rich mixtures, the formation of COS is still in progress after 6 milliseconds.

12) In fuel-rich mixtures the concentration of NO attains a maximum value in about 0.3 ms and thereafter decreases slowly.

13) In fuel-lean mixtures the concentration of NO rises rapidly immediately behind the flame front and then slowly thereafter.

14) The effect of fuel-sulfur on NO formation is apparently confined to the immediate post-flame region.

15) The experimental observation that fuel-sulfur and fuel-nitrogen do not significantly influence the CO concentration in the burned gas was confirmed by the theoretical calculations.

Since this work was completed, Sarofim and Beér (16) and de Soete (17) have presented experimental evidence that NO may be destroyed on alumina surfaces under fuel-rich conditions in the presence of CO. This possible effect was not tested in our experimental work or taken into account of in our theoretical model. It may have been a factor in producing the measured values which are shown in Figures 1 to 6. The omission of this mechanism may accordingly lead to error in the predictions of our model for fuel-rich mixtures. However, the previously cited agreement in references (8) and (9) of the predictions of this model with experimental measurements for thermal and fuel-NO_x, does not indicate any such error, at least for equivalence ratios less than 1.3. It may be noted that our wall temperatures were much higher than those of references (16) and (17), and also that the high water content of our burned gas may have poisoned the alumina wall. In any event the detailed predictions of our model must be considered tentative for $1 < \phi < 1.1$-1.3 and only speculative for $\phi > 1.3$.

NOTATION

ppm: parts per million (molar basis)

ϕ: equivalence ratio $= \dfrac{\text{fuel-to-air ratio}}{\text{fuel-to-air ratio for stoichiometric mixture}}$

LITERATURE CITED

1. Chen, J.L.-P., and Churchill, S.W., Combust. Flame, 18, p. 37, (1972).

2. Chen, J.L.-P., and Churchill, S.W., Combust. Flame, 18, p. 27, (1972).

3. Bernstein, M.H., and Churchill, S.W., Sixteenth Symposium (International) on Combustion, p. 1737, The Combustion Institute, Pittsburgh, (1977).

4. Choi, B., and Churchill, S.W., Advances in Chemistry Series No. 166, Evaporation - Combustion of Fuels (J.T. Zung, Ed.), p. 83, American Chemical Society, Washington, DC, (1978).

5. Choi, B., and Churchill, S.W., Seventeenth Symposium (International) on Combustion, p. 917, The Combustion Institute, Pittsburgh, (1979).

6. Goepp, J.W., Tang, H.S.K., Lior, N., and Churchill, S.W., AIChE Journal, 26, 855 (1980).

7. Tang, S.-K., and Churchill, S.W., Eighteenth Symposium (International) on Combustion, p. 73, The Combustion Institute, Pittsburgh, (1981)

8. Tang, S.-K., and Churchill, S.W., Chem. Eng. Comm., 9, 137 (1980)

9. Tang, S.-K., and Churchill, S.W., Chem. Eng. Comm., 9, 151, (1980)

10. Wendt, J.O.L., and Ekmann, J.M., Combust. Flame, 25, p. 355, (1975).

11. Wendt, J.O.L., Morcom, J.T., and Corley, T.L., Seventeenth Symposium (International) on Combustion, p. 671, The Combustion Institute, Pittsburgh, (1979).

12. de Soete, G., La Formation des oxyde d'azote dans la zone d'oxidation des flammes d'hydorcarbures, Institut Francais du Petrole, Final Report No. 23306, Ruielle Malmaison, France (June 1955).

13. Johnson, G.M., Smith, M.Y., and Mulcahy, M.F.R., Seventeenth Symposium (International) on Combustion, p. 647, The Combustion Institute, Pittsburgh, (1979).

14. JANAF Thermochemical Tables, Dow Chemical Co., Midland, Michigan, (December 31, 1960 to June 30, 1970).

15. Tang, S.-K., The Formation of NO_x from Combustion in a Refractory Plug Flow Burner of Hexane Droplets Doped with Nitrogen and Sulfur Compounds, Ph.D. Thesis, University of Pennsylvania (1980).

16. Sarofim,A.F., and Beer,J.M., Eighteenth Symposium (International) on Combustion, p. 111 , The Combustion Institute, Pittsburgh (1981).

17. de Soete, G., Proceedings Sixth Members Conference, IFRF, Noordwykerhout, (May 1980).

NUCLEATION/GROWTH RATE KINETICS OF GYPSUM IN SIMULATED FGD LIQUORS: SOME PROCESS CONFIGURATIONS FOR INCREASING PARTICLE SIZE

Growth and nucleation rates for Calcium Sulfate dihydrate (Gypsum) in simulated flue-gas desulfurization (FGD) liquor were experimentally determined and were used in a computer model to predict the product size from several crystallizer configurations. A crystallizer design which accelerates the removal of small crystals increases the product size by a factor of two over the current precipitator design.

DAVID L. ETHERTON
and
ALAN D. RANDOLPH

University of Arizona
Tucson, Arizona

Development of effective and reliable methods for removal of SO_2 from stack gases of coal fired plants is important to the nations energy and environmental future. This task looms especially important in light of recent mandates to reduce reliance on imported oil by shifting to coal-fired power plants. Many flue gas desulfurization (FGD) processes have been developed over the last 20 years and several processes are successfully being used in full-scale operations.

The most common process currently on-line is the lime-limestone scrubber. In this process SO_2 is absorbed by a slurry of lime or limestone which reacts to form calcium sulfite or, if excess air is supplied, calcium sulfate. The sulfite/sulfate is precipitated as a fine solid and is disposed of as land-fill. Figure 1 shows a typical scrubber configuration.

One of the problems with this process is that the solids formed are relatively small, requiring large clarifiers and filters to remove enough water from the sludge to make an acceptable land-fill. The purpose of this study was to determine crystallization kinetics of the calcium sulfate dihydrate (gypsum)

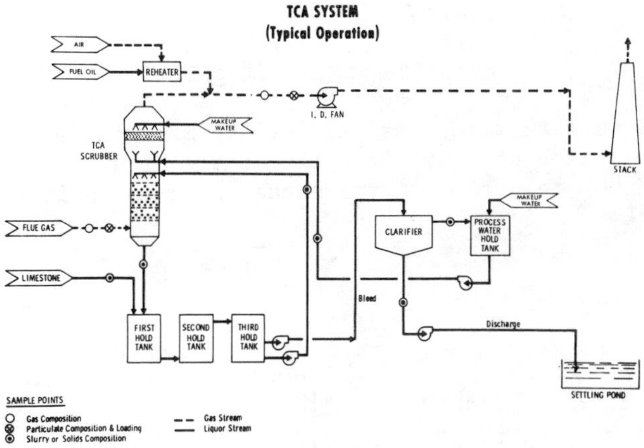

Figure 1. Flow diagram of typical scrubber.

system and to use this kinetic information in a computerized crystal size distribution (CSD) simulator to predict conditions which might increase particle size. Growth and nucleation kinetics were determined by examining the effects of supersaturation, slurry density, pH and organic additives, on growth and nucleation rates. These data were then combined to give empirical correlations which were used in the CSD simulation to study alternate crystallizer configurations.

APPARATUS

The kinetic experiments were carried out in a "minucleator" apparatus consisting of a

---

D.L. Etherton is now with David Taylor Naval Ship Research and Development Center, Bethesda, MD.

0065-8812-81-4578-0211-$2.00

crystallizer with associated feed tank, feed pump, filters, and particle counter (Figure 2). The crystallizer was a 1-liter jacketed glass vessel which was clamped to a machined plexiglass base. A 150 μm stainless steel screen was placed over the exit stream port which provided complete retention of the large (> 150 μm) seed crystals needed to provide secondary nucleation but allowed the nuclei (< 50 μm) to pass to a particle counter. The particle counter was a Particle Data Inc., counter which uses the zone sensing principle to count and size particles. Growth and nucleation rates were calculated from particle counts with an associated PDP-8 mini-computer. This mininucleator system is described in more detail by Cise and Randolph (1) and Shadman and Randolph (2).

The feed liquor was a saturated solution of calcium sulfate with ions added to simulate actual stack-gas liquor as follows:

calcium	1.79 g/L
magnesium	0.20 g/L
sodium	0.05 g/L
sulfate	1.79 g/L
chloride	2.51 g/L

Supersaturation was attained by metering a solution of K_2SO_4 into the crystallizer. The K_2SO_4 reacted with the excess calcium to form calcium sulfate. Thus the experimental system symulated a fully oxidized FGD scrubber.

DATA ANALYSIS

Data analysis consisted of determining growth and nucleation rates from particle count data and measuring or calculating supersaturation and seed crystal slurry density. The particle size-number data was analyzed by the well known population balance. On-line data regression to Equation (1) was performed by the minicomputer within a period of a few seconds after data measurement.

$$n(L) = \frac{B^o}{G}\exp\left(\frac{-L}{G\tau}\right) \qquad (1)$$

Figure 3 shows typical information as output from the PDP-8 minicomputer. The growth rates of the retained seed crystals were calculated from measured size and mass gains during a run.

The solubility of a multi-component ionic system is difficult to define due to the interaction of ionic and solid-liquid equilbria of the various species. A solution equilbria computer program, the Bechtel Modified Radian Equilbrium Program (EPA, 3) has already been developed to cal-

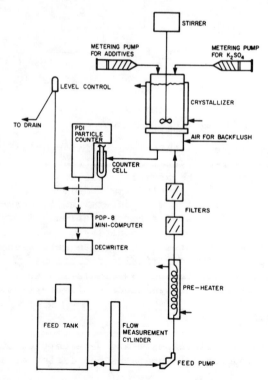

Figure 2. Schematic flow diagram of experimental apparatus.

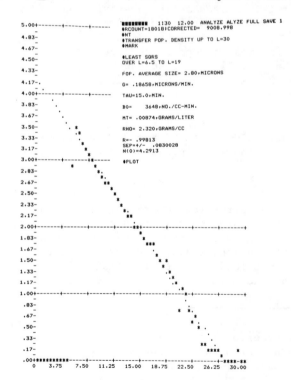

Figure 3. Typical data output from PDI counter.

culate equilibrium compositions in stack-gas liquors and was used to calculate the supersaturation of calcium sulfate used in this study. Throughout this paper supersaturation is defined as the amount of solute in solution in excess of the saturation concentration, expressed as g/L of $CaSO_4 \cdot 2H_2O$.

The nucleation rate was correlated with seed growth rate and the slurry density of the seed crystal in the following power-law model.

$$B^O = k_n G_s^i M_T^j \qquad (2)$$

Growth Rate was correlated with supersaturation as

$$G_s = k_g S^a \qquad (3)$$

Similar power law correlations are in general use in the crystallization literature.

EXPERIMENTAL RESULTS

Nucleation and Growth Kinetics

Calcium sulfate was shown to nucleate by secondary mechanisms in the presence of 150 to 300 μm seed crystals and at supersaturations less than 2.5 g/L. However, at higher supersaturations or without seed crystals bursts of primary nucleation would occur.

Secondary nucleation was correlated with seed growth rate and seed slurry density to give the kinetics expression:

$$B^O = \exp(16.72) \; G_s^{1.48} M_T^{1.27}$$

with a regression coefficient (R^2) of 0.83. Figure 4 is a plot of these data and the correlation line.

An emperical growth rate correlation is given as

$$G_s = \exp(13.11) \; S^{2.23}$$

with a regression coefficient (R^2) of 0.95. These data and the correlation line are shown in figure 5.

Effects of pH Level

pH is an important design variable in FGD scrubbers, which is used to control both solution and solid equilibria. The pH also varies in different parts of the scrubber

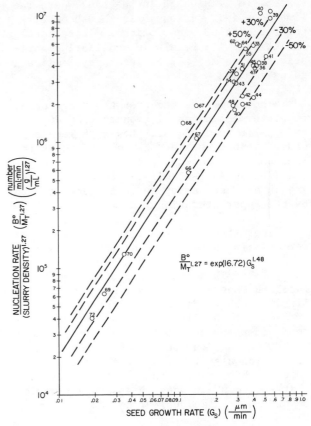

Figure 4. Correlation of gypsum nucleation rates.

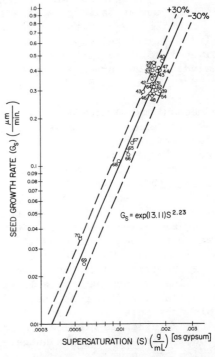

Figure 5. Growth rate correlation for gypsum.

system, (e.g. high pH in the tank where the lime is added and low pH in the absorber tower due to SO_2 absorption). These local pH variations will result in equilibrium variations which may affect local crystallization rates.

The pH was varied over the range of 3.0 to 7.8 by adding small amounts of H_2SO_4 to the pH 7.8 feed liquor. Lower pH levels caused both growth and nucleation rates to increase. Since the pH affects solution equilibrium and solubility its effect on growth and nucleation rates is accounted for by the supersaturation.

The effects of localized pH variations were simulated by adding several drops of H_2SO_4 to a steady-state run having an initial pH of 7.8. This experiment simulated the local pH changes that might occur in a poorly mixed scrubber system. Nucleation rate increased 2 to 5 times the steady-state value soon after the pH was lowered. As the acid washed out and the pH returned to normal, nucleation also returned to the original value. This simple experiment illustrates that sudden pH changes in a scrubber would result in erratic nucleation rates.

Effects of Additives

The use of additives to modify nucleation and growth rates is a common method for improving CSD in industrial crystallizers. Additives can also be used to change crystal habit. It is therefore, expected that the use of additives in the crystallization of calcium sulfate sludges might produce crystals of larger size or better shape.

One mechanism by which an additive operates is by adsorbing on the crystal surface. This surface modification can cause both growth and nucleation rates to be modified. The additive may adsorb on a particular face of the crystal thus causing the crystal to grow to a different shape (Michaels et al., 4). An additive may also modify the crystal habit by providing sites which promote heterogeneous nucleation (e.g., long-chain organic acids, Sarig and Ginio, 5). Such additives have been used in boilers and cooling towers to prevent scaling by forcing crystallization in the solution rather than on the walls. It is possible that this type of additive could be used in a scrubber to provide a constant level of nucleation rather than unpredictable cycling which can occur.

The additives chosen for this study have been used in stack-gas systems or have been used in other industrial environments. Sodium dodecyl benzene sulfonate (SDBS) has been reported to be a growth and nucleation modifier for other systems, e.g. Randolph and Puri (6). Calgon CL246, a mixture of poly-acrylic acids, is used as a scale preventative in cooling towers. Adipic acid is added to lime/limestone scrubbers as a pH buffer. Citric acid was shown to be an effective crystal habit modifier of calcium sulfate by McCall and Tadros (7).

The results of the experiment with additives are presented in Table 1. The ideal additive would decrease nucleation, increase growth, and change the crystal habit to a blockier shape. None of the additives studied performed in this manner. However, both SDBS and CL246 act to promote heterogeneous nucleation at constant levels. This study also confirms the findings of McCall and Tadros that show citric acid to change the crystal habit. Figure 6 shows crystals formed with (c) and without (a and b) citric acid additives. (The large crystals in (a) are seed crystals). The change from long needles to blockier crystals would provide a sludge that was easier to filter.

Table 1   Effects of Additives on Crystallization Parameters

Additive	Concentration	EFFECT ON		
		Nucleation Rate	Growth Rate	Crystal Habit
SDBS	10-40 mg/L	increased	none	none
Calgon CL246	100-500 mg/L	increased	none	none
Adipic Acid	1.0-6.0 g/L	none	none	none
Citric Acid	1.0 g/L	none	decreased	blockier crystals

CSD MODELLING

Gypsum growth and nucleation kinetics can be used in conjunction with mass and population balances and given design parameters (e.g., vessel volume, feed supersaturation, liquid and solid residence times) to predict CSD of the calcium sulfate product from a scrubber precipitator. A computer simulation program (Mark I CSD Simulator) has been developed previously (Nuttall, 8), to predict CSD from various crystallizer designs. This program was used to study the effect on CSD of varying liquid vis-a-vis solids residence times and to examine the effect of high levels of nucleation which might be caused by localized low pH and/or additives which enhance heterogeneous nucleation.

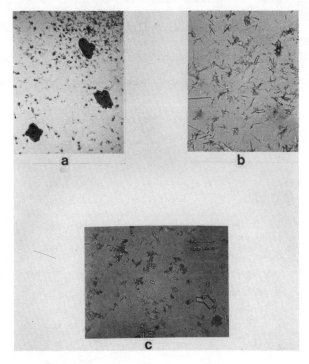

Figure 6. Gypsum Crystals: (a) agglomerated needles, (b) single crystal needles, (c) crystals grown in citric acid additive.

Simulation Procedures

The Mark I simulator simultaneously solves, population and mass balances, growth kinetics, and nucleation kinetics to give the steady-state CSD of the crystallizer product (e.g. Randolph and Larson, 9, Chap. 8). This program can simulate single- or multi-stage crystallizers with or without product classification and fines destruction. In this study Mark I was used to simulate a single-stage gypsum crystallizer with and without classified product removal.

Classified product removal (in this case preferential retention of larger-sized product) might be accomplished internally or by installing some type of classification device to the exit stream. One stream of the classification device is returned to the crystallizer and the other is removed. The result of such classification would be to cause the crystals to have a size-dependent residence time. Proper selection of the size-dependence would produce larger crystals.

Three configurations with different liquid-solid residence times were modeled, (a) mixed suspension mixed product removal (MSMPR), (b) the current configuration of the

Shawnee facility, and (c) the proposed design. Figure 7 shows the schematic of each configuration and the form of the classification function, which is the ratio of the flow rate of particles in a given size range to the flow rate in the mixed removal stream.

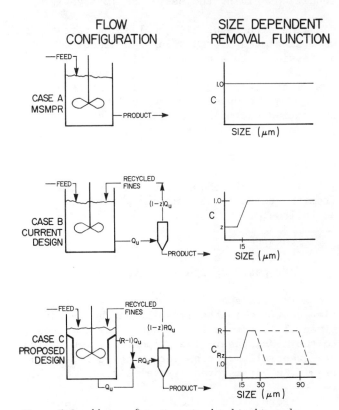

Figure 7. Scrubber configurations simulated in this study.

For the MSMPR configuration the liquid and solid residence times are equal, therefore, the classification function is constant for all sizes. In the current Shawnee design the clarifier overflow stream contains particles smaller than 15 μm. This results in crystals smaller than 15 μm being removed at a lower rate than those in a mixed suspension. The proposed design would incorporate a baffle-type settler inside the crystallizer. This baffle would provide a zone of low mixing where large particles would settle back into the mixing zone and small particles would be removed. The larger particles would thus be removed at a reduced rate. However, the clarifier overflow would still return particles less than 15 μm to the crystallizer (this is a design constraint on the clarifier) and they would thus be removed at a lower rate than particles between 15 μm and the upper classification size. The upper classification size and the removal ratios are variables which depend on particle hydraulics and settler design. Classification sizes of 30 to 70 μm and removal ratios of 3 to 10 were simulated in order to determine the conditions giving the best size improvement.

The effects of high nucleation rates were simulated by replacing the kinetic expression with a constant value of nucleation. This simulates the nucleation rate that might be produced by certain additives and gives the upper asymptotic limit of nucleation caused by localized zones of low pH as measured in this study.

Results of CSD Simulations

The calculated product sizes resulting from the three configurations simulated are as follows:

MSMPR	71 μm
Current design	67 μm
Proposed design	83 to 130 μm

These results show that returning the clarifier overflow to the crystallizer has a detrimental effect on the crystal size compared to the MSMPR case. The proposed design, which preferentially retains larger crystals, would result in an improved crystal size.

Figure 8 shows the size of the crystals in the product (a) and in the crystallizer (b) as a function of the classification size and the removal ratio. The product size increases with increasing removal ratio and decreasing classification size to 50 μm, then levels off or decreases. The size of particles in the crystallizer also increases with increasing removal ratio, but reaches a maximum as a function of classification size. The classification size at which the maximum occurs increases with increasing removal ratio.

From Figure 8 it appears that a good design would be a classification size of 50 μm and a removal ratio of 7. This design gives a product size of 117 μm and a size in the crystallizer of 180 μm. Thus, the product size has increased by a factor of two over the current design. Such classified retention should be readily achievable in the gypsum-water system.

In order to study the effect of high nucleation rate, the kinetic expression for nucleation rate was replaced by a constant

value. This was done for the MSMPR, the current design, and the optimum proposed design cases using nucleation rates of 4000, 8000, and 20000 number/mL min. The product size for each of these cases is given in Table 2. The nucleation rate calculated using the experimental correlation was between 1000 and 2500 for these three cases.

Table 2   Effects of High Nucleation Rates on Product Size

Nucleation Rate	MSMPR (μm)	Current Design (μm)	Proposed Optimum Design (μm)
Experimental Kinetics	71	67	117
4,000	48	39.8	77
8,000	38	30.4	29.7
20,000	28	21.6	16.1

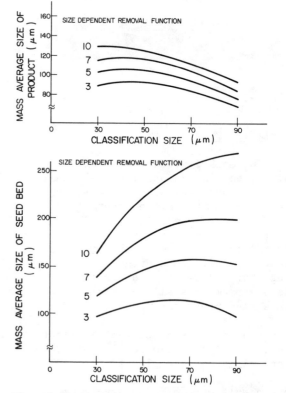

Figure 8. Mass average crystal size as a function of classification size and removal ratio; (a) product, (b) suspension.

Increasing the nucleation rate causes significant reduction of product size in all cases, and in fact, ultimately invalidates the benefits of the proposed design. Nucleation rates of 4000 to 8000 would be expected in areas of localized low pH. Nucleation rates of 8000 to 20000 occur in the presence of heterogeneous nucleation enhancers. It is desirable to eliminate both of these sources of nuclei if the particle size is to be maximized.

The proposed crystallizer modification may result in some undesirable changes in the scrubber operation. Because of the increased particle size and the accelerated removal of fines, the area of crystals in the crystallizer might be reduced. (This would depend on whether or not total solids concentration increased). Reduced crystal area could cause increased fouling on the scrubber internals, as fouling is known to occur if sufficient crystal area is not provided in the scrubber. The crystal area is dependent on the classification size and the removal ratio just as is the product size. Therefore, some trade-offs may be required between increased particle size and increased tendency to foul, if further increases in slurry density are not acceptable. Finally, the proposed design would entail larger liquor flows to the settler/classifier in order to keep total solids content in bounds in the scrubber/crystallizer, although total solids loading of fines would decrease. If the settler is liquid-phase limited the additional volumetric loading would require a larger settler/classifier.

SUMMARY AND CONCLUSIONS

1. Nucleation and growth rates of calcium sulfate dihydrate (gypsum) crystals were measured in simulated stack-gas liquors as a function of supersaturation, pH, slurry density, and chemical additives. These data were correlated to give design-useful kinetic expressions. The effect of additives on crystal habit was also observed.

2. From the study of the effects of supersaturation and slurry density it was shown that gypsum nucleates by secondary nucleation mechanisms when crystals greater than 150 μm are retained in the slurry. Bursts of primary nucleation, which decrease particle size, will occur if the supersaturation becomes too high or if seed crystals are not present.

3. Low pH was found to increase both growth and nucleation rates. Regions of low pH (or sudden decreases in pH) produced bursts of nucleation resulting in smaller crystals.

4. Additives studied were sodium dodecyl benzene sulfonate, polyacrylates, adipic acid, and citric acid. Of these, only citric acid improved the crystal habit by producing block-ier crystals rather than the needle-like crystals normally produced.

5. A crystallizer design that would accelerate the removal of small crystals relative to larger ones was shown to produce a larger particle size. This type of crystallizer is common in industry and the resulting size improvement over the MSMPR crystallizer is well-known. The key to successful operation of this type of crystallizer is the ability to control, or at least to predict, the nucleation rate. This is best done by operating under conditions where secondary nucleation is the predominate source of nuclei. Crystallization modelling should be combined with scrubber operating experience to provide an integrated design of the proposed precipitator configuration.

ACKNOWLEDGEMENT

The authors are indebted to the Electric Power Research Institute, Inc. (EPRI) for financial support under contract number RP1030-2. The following legal notice is required by EPRI.

This work was prepared by the University of Arizona as an account of work sponsored by EPRI. Neither EPRI, members of EPRI, nor the University of Arizona, nor any person acting on behalf of either:

a. Makes any warranty or representation, express or implied, with respect to the accuracy, completeness, or usefullness of the information contained in this report, or that the use of any information, apparatus, method, or process disclosed in this report may not infringe on privately owned rights; or

b. Assume any liabilities with respect to the use of, or for damages resulting from the use of, any information, apparatus, method, or process disclosed in this report.

NOTATION

a = emperical supersaturation power

B^o = nucleation rate, number/mL min

G = growth rate, μm/min

G_s = seed growth rate, μm/min

i = emperical seed growth power

j = emperical slurry density power

k_g = growth rate power law rate constant

k_n = nucleation power law rate constant

L = crystal size, μm

M_T = slurry density, g crystals/mL liquor

n(L) = population density at size L, number /mL·μm

R^2 = regression coefficient

S = supersaturation, g/L

τ = residence time, min

LITERATURE CITED

1. Cise, M.D. and A.D. Randolph, "Secondary Nucleation of Potassium Sulfate in a Continuous-Flow, Seeded Crystallizer," presented at Third Joint Meeting, AIChE-IMIQ, Denver (Sept. 1970).

2. Shadman, F. and A.D. Randolph, AIChE J., 24 (5), 782 (1978).

3. Environmental Protection Agency, Report EPA-65012-75-047, "EPA Alkalic Scrubbing Test Facility: Summary of Testing Through October 1974" (June 1975).

4. Michaels, A.S., P.L.T. Brian and W.F. Beck, "Cinemicrographic Studies of single Crystal Growth: Effects of Surface Active Agents on Adipic Acid Crystals Grown from Aqueous Solution," presented at the 57th annual meeting, AIChE (1964).

5. Sarig, S. and O. Ginio, J.Phy. Chem., 80 (3), 256 (1976).

6. Randolph, A.D. and A.D. Puri, AIChE J., 27 (1), 92 (Jan. 1981).

7. McCall, T., and M.E. Tadros, "Effects of Additives on Morphology of Precipitated Calcium Sulfate and Calcium Sulfite - Implications on Slurry Properties," unpublished manuscript, Martin Marietta Corp., Baltimore (1979).

8. Nuttall, H.E., "Computer Simulation of Steady State and Dynamic Crystallizers," Ph.D. Dissertation, University of Arizona, Tucson (1971).

9. Randolph, A.D. and M.A. Larson, Theory of Particulate Processes, Academic Press, New York (1971).

MONOGRAPH SERIES